The ABC of Armageddon

The ABC of Armageddon

Bertrand Russell on Science, Religion, and the Next War, 1919–1938

Peter H. Denton

STATE UNIVERSITY OF NEW YORK PRESS

Published by
State University of New York Press, Albany

© 2001 State University of New York

All rights reserved

Printed in the United States of America

No part of this book may be used or reproduced in any manner whatsoever without written permission. No part of this book may be stored in a retrieval system or transmitted in any form or by any means including electronic, electrostatic, magnetic tape, mechanical, photocopying, recording, or otherwise without the prior permission in writing of the publisher.

For information, address
State University of New York Press,
90 State Street, Suite 700, Albany, NY 12207

Production by Michael Haggett
Marketing by Jennifer Giovani

Library of Congress Cataloging-in-Publication Data

Denton, Peter H., 1959–
 The ABC of Armageddon: Bertrand Russell on science, religion, and the next war, 1919–1938 / Peter H. Denton.
   p. cm. – (SUNY series in science, technology, and society)
 Includes bibliographical references (p.) and index.
   ISBN 0-7914-5073-2 (alk. paper)–ISBN 0-7914-5074-0 (pbk. : alk. paper)
   1. Russell, Bertrand, 1872–1970. 2. Science—Philosophy—History— 20th century. 3. Religion and science—History— 20th century. 4. Social ethics—History— 20th century. I. Title. II. Series.

B1649.R94 D36 2001
192—dc21   2001031189

10 9 8 7 6 5 4 3 2 1

Contents

	Preface	vii
	Introduction	xi
Chapter 1	Science and the New Civilization	1
Chapter 2	The Individual and the Machine	18
Chapter 3	Religion, Metaphysics, and Meaning	41
Chapter 4	Physics and Philosophy	60
Chapter 5	Science, Religion, and Reality	83
Chapter 6	Science and Power	110
	Notes	127
	Bibliography	161
	Index	171

Preface

Centuries are bridged by the lives of people who carry into the new era the ideas and opinions of the old. Born on May 18, 1872, into the aristocracy of British liberalism, the son of John Russell, Viscount Amberley, and his wife Katherine Stanley, Bertrand Russell had an ascetic upbringing at the hands of his paternal grandmother after he was orphaned at age three. He became one of the foremost mathematicians and philosophers of the twentieth century almost from its beginning, publishing *Principia Mathematica* with Alfred North Whitehead in 1910–1913.

Yet it is his appeal as a writer and a thinker outside the walls of the university that is the most intriguing, earning the third Earl Russell, among other honors, the Nobel Prize for Literature in 1950. He wrote dozens of books, hundreds of articles, and thousands of letters in the ninety-eight years of his life. After the devastation of the Great War of 1914–1918, many of these dealt with the future direction of Western society and the consequences of the science and technology that powered industrial civilization in all of its political guises. Russell brought a liberal view of the individual and of society into a new century whose events came to threaten the one just as much as the other. Cofounder of the Pugwash movement and outspoken leader in the midst of the early Cold War of the opposition to nuclear weaponry, he was accustomed to matching his actions to his opinions on everything from politics to disarmament, from education to economics, and from the nature of happiness to the notorious conduct of his marital affairs.

This book covers a crucial period both in the history of twentieth-century Western culture and in Bertrand Russell's life, a period in which he and his Anglo–American contemporaries struggled to articulate what kind of a world was possible, and what kind of a world they feared, in the aftermath of the Great War. Science bulked large in their vision of what was to come, and the technology that had shown its ugly power in the weapons of modern war could never again be viewed with the same optimism with which they had greeted the dawn of this new century in 1900.

It was a conversation more than an intellectual movement, a conversation carried on in full view of a world reeling from the effects of the Great War and the opportunities and dangers its conclusion presented. It was a conversation in which the public itself participated, as reader, audience, and author in its own right, through periodicals and other popular forms of print and the new venue of radio broadcasting.

By focusing on Russell's largely neglected contribution to this public conversation and by introducing some of the ideas to which he was responding, I hope to contribute to a similar conversation at the start of our century. The same issues and concerns still confront us and, if anything, in the last seventy-five years we have only managed to muddle the questions that Russell and his contemporaries articulated much more clearly. The apocalyptic Next War to end civilization continues to haunt our society, as does the increasing problem of a guerrilla warfare with the ecology of the planet which, despite our science and technology, we can never expect to win.

And so, in the midst of the Information Age, in the face of the ongoing economic and technological miracles of industrial civilization, in the bright light of our scientific view of the universe, the same fundamental problems remain: it is not our knowledge that needs to increase but our understanding of what it means; it is not our science that needs to grow and mature but our view of the individual and society; it is not our tools that need development but the judgment and wisdom with which they are used to create whatever future will follow.

To understand the prospect of impending Armageddon—as an end of our own making—we need to lay out the elements of its construction in simple and transparent terms, to communicate them outside of whatever circle of experts to the people who can appreciate the consequences to their own lives of the decisions which others must make. An ABC of Armageddon—like an ABC of relativity to explain the new physics or an ABC used to teach reading and writing—has its value not as the final word but as an introduction, encouraging individuals to read, learn, and act as they choose toward the future that they desire for themselves and their children.

What Russell and others had to say on the subjects that follow continues to be relevant today.

In the trajectory of this work, I have incurred a range of intellectual and personal debts, for which what follows is offered as small compensation:

To Louis Greenspan, my friend and dissertation supervisor—an order established from the outset—I owe the tolerance, wisdom, and good humor that allowed me to take an interdisciplinary path unusual even in the institutional context of a department of Religious Studies.

To my friends, colleagues, and students, I owe the stimulation and correction—often as steel to my flint—that sparked the ideas which this book and my other work contains, though responsibility for their final form remains my own.

To my family, which has long humored me for my interest in Russell and his cohorts, I owe the love and support that has made this work possible. In an age of voluminous speech, the significance of a thought is still not measured by the number of words in which it is expressed, and so this book is dedicated with love to Mona and our children, Ruth and Daniel, in thanksgiving for all the cup of life contains.

Finally, for help in bringing this work into its public form, I am grateful to Jim Farris, Kevin Farris, Steve Fuller, Geoffrey Cantor, the editorial staff of State University of New York Press, and Sal Restivo, who would likely prefer to be considered its godfather rather than its midwife, but who was both.

Winnipeg
September 2000

Introduction

As we enter a new century and a new millennium, it is hard for us to comprehend the emotional and psychological aftermath of the Great War. The post-War period tends to be described in terms of the political, social, and economic changes that distinguished the world of 1919 from that of 1914, or by lists of the changes wrought by science and technology that, in their turn, would shape a different society. Yet the effects of observing a civilization in crisis, of watching political institutions crumble, of experiencing the slaughter of soldiers and civilians on a scale never seen before in modern Europe, or the profound uncertainty of what would be left if and when the fighting stopped, scarred the generations whose lives were disrupted by the Great War. We read it in their books, their letters, their diaries, but we find it difficult to understand a perspective in which our own existence was only one prospect among many in 1919, and perhaps not the most likely. What seems to us, in hindsight, an inevitable development in Western society was to them only a possibility; they had seen the "war to end wars" turn into a war that threatened to end the world as they knew it.

Thus while the intellectual concerns of Bertrand Russell and his Anglo-American contemporaries were indisputably shaped by the Great War, their discussions of the future of modern society took place within the context supplied not only by the consequences of one war but also by the prospect of another, more devastating conflict. There was a general feeling of apprehension that made "the Next War" a psychological reality long before the outbreak of war in 1939.¹ It was the need to change the conditions that seemed to make another such conflict unavoidable which gave a special urgency and pragmatic cast to numerous discussions about the emerging scientific society in what I have termed "the interwar period." For those who lived in the midst of it, the interwar period was not the chronological period between the First World War and Second World War but rather the indefinite psychological moment between the Great War and the Next War, between the horror of what they had experienced, and the dread of the Armageddon that the Great War foreshadowed.

The interwar period also constitutes a more distinct period in the work of Bertrand Russell than has previously been recognized. Perhaps the most unequivocal of Russell's various observations that the Great War marked a personal watershed for him is found in *Portraits from Memory:* "My life has been divided into two periods, one before and one after the outbreak of the First World War, which shook me out of many prejudices and made me think afresh on a number of fundamental questions."² Russell went on to say that although he "did not completely abandon logic and abstract philosophy," he became "more and more absorbed in social questions and especially in the causes of war and the possible ways of preventing it."³ He did not regard the human questions produced by the Great War as dispassionately as he had earlier considered the principles of mathematics and logic. He said that the war made it "impossible" for him "to go on living in a world of abstraction," for he felt "an aching compassion" for the young men he saw boarding troop trains "to be slaughtered on the Somme because generals were stupid."⁴ He said he had become "united to the actual world in a strange marriage of pain" and that, because of the War, he had lost "the hope of finding perfection and finality and certainty."⁵

> All the high-flown thoughts that I had had about the abstract world of ideas seemed to me thin and rather trivial in view of the vast suffering that surrounded me. The non-human world remained as an occasional refuge, but not as a country in which to build one's permanent habitation.⁶

What it was possible for human beings to know about themselves or anything else thus had become less of an abstract epistemological problem for Russell and more of a practical and an emotional one, with important social consequences.

It was this movement away from the abstractions of mathematical logic and toward social and political questions that in *My Philosophical Development* Russell dubbed his "retreat from Pythagoras."[7] Though his retreat began early in the century, he said the "ascetic mood" that inspired his mystical view of mathematics had been "finally dispelled by the First World War."[8] Pythagoras functioned for Russell as a metaphor of the abstract, intellectual truth about the universe, which before the War he thought could be expressed in mathematical terms.[9] Russell's "retreat from Pythagoras," from the certainties of abstract mathematical truth to the uncertainties of the contingent world of social power relations, from the abstract notion of "the Good" toward the practical problems of its realization in society, thus characterizes the pragmatic character that his writing tended to reflect during the interwar period. Whatever his concern with philosophical questions about the nature of meaning and truth, after 1918, Russell also seemed compelled to address the practical question of "what ought we to do?" in a way that took him into the heart of the problem of formulating a modern social ethic and thus trying to prevent the Next War. He undertook the task of analyzing and writing about what had happened, and what science and technology meant for post–War industrial civilization, in the hope that a popular primer of impending disaster—an ABC of Armageddon—might forestall what seemed, at times, to be an inexorable outcome.

Without certain very definite actions, such as the elimination of nationalism and a more equitable distribution of wealth and power, he believed that the Next War seemed unavoidable:

> Material progress has increased men's power of injuring one another, and there has been no correlative moral progress. Until men realize that warfare, which was once a pleasant pastime, has now become race suicide, until they realize that the indulgence of hatred makes social life impossible with modern powers of destruction, there can be no hope for the world. It is moral progress that is needed; men must learn toleration and the avoidance of violence, or civilization must perish in universal degradation and misery.[10]

Of all the major factors he identified as contributing to the likelihood of the Next War, however, the most compelling problem for

Russell was that presented by the "mechanistic outlook" of contemporary society. He cast the problem of "the old savage" in terms of the struggle between humanity and industrialism, or between the individual and the machine. The survival of civilization through socialism and internationalism might be accomplished by the use of sufficient force, but Russell was concerned about the kind of society that would then result. He preferred to achieve the moral progress necessary for the survival of the old savage through a change in the outlook of individuals and fulminated against the mechanistic and utilitarian outlook he currently observed, in which science was understood primarily in terms of its applications.

Though he retreated from the mysticism of Pythagoras after the Great War, Russell did not become a disciple of Henry Ford. The misrepresentation of science as nothing more than technique, in which the value of knowledge was measured not by its truth but by its utility, aroused Russell's ire throughout the interwar period. He believed that industrial civilization was concentrating too much on scientific technique, or the applications of science, and was not recognizing the importance of the scientific outlook that had led to these applications in the first place. Alarmed by the current trend toward what he termed *instrumentalism,* he maintained that science involved more than the instrumental epistemology he perceived in the application of scientific technique to the problems of Western society.

Realizing the inevitability of industrial development, and what its consequences would be in terms of the organization, interdependence, and structure of life in the Machine Age, Russell acknowledged the global character of an industrial economy, with its large-scale production and distribution of commodities, and the need for efficient organization which this required. He also saw the dangers posed for the future of industrial civilization as a whole by the manipulation of such a global economy through the competing interests of national states. Like many others, he saw that increased social organization inevitably came at the expense of individual freedoms and tended toward authoritarian rule, where the interests of any individual could be oppressed by the collective will of the majority. He grappled with the conundrum posed by science freeing people from the constraints of nature while further constraining individual freedoms by the nature of society that industrialism made necessary. He also realized how vulnerable such a society would be to manipulation through propaganda, especially if public debate were not encouraged or dissent not tolerated.

POPULAR PHILOSOPHY AS PERFORMANCE

There are several implications that follow from considering what Bertrand Russell wrote between the Great War and the Next War. First, it is important to place Russell within the context of the popular Anglo-American conversation on interests similar to his own. His ideas were not entirely original and bear considerable resemblance to what other people were expressing at the time. He was a reactive writer, superbly suited by his education and writing ability to engage in conversation on a wide range of subjects, and he was able to represent other people's ideas so effectively that they often seemed to be his own. To understand what he wrote, therefore, requires us first to understand what other people were writing at the same time. Second, because this was a "public conversation" rather than simply a debate among scholars or some kind of intellectual elite, Russell's work was aimed as much at persuading the public of the validity of his position as it was at undermining the ideas with which he disagreed. Thus there is a strong element of "performance" in much of what he wrote during the interwar period, making the philosophical inconsistencies of an argument of less immediate concern to him than its persuasiveness.

One of the developments in Russell's writing after 1900, which may be linked to his progressive "retreat from Pythagoras," was his shift toward writing less technical articles and books, aimed at a popular audience. The commercial success of his "shilling shocker," *The Problems of Philosophy* (1912), written for the Home University Library, brought home to Russell the impact that such writing could have on the popular understanding of issues, and thus on public opinion. This shift became more pronounced after 1918. In his *Autobiography,* he painted the effects of the Great War on his own life with broad brush strokes. It "changed everything" for him; he ceased to be "academic" and took to writing "a new kind of books [sic]."[11] In *Religion and Science* (1935), another volume written for the Home University Library, Russell observed that there were two means of influencing ethical choices: the way of the legislator and the way of the preacher. For example, while the legislator had the instruments of the State at his disposal to compel assent, the preacher had only the emotional persuasiveness of his presentation to effect the same result. Having failed in his attempts to become a legislator, throughout the interwar period Russell developed and expressed his ethical ideas as a "preacher," thus making both his delivery and the expectations of his intended audience into crucial components of what he wrote.

Russell wrote hundreds of articles and books for the popular press during the interwar period, and his writing style definitely could be termed *sermonic*.[12] His published work from the interwar period therefore should be considered as much in terms of its rhetorical performance as its content, philosophical or otherwise. His stance as a "preacher" determined to a large extent the way in which he participated in the public conversation on religion, science, and social ethics during that time. To focus on the philosophical inconsistencies in his ethical writings, for example, would be to miss the point of his "sermons." Each one had its own intended audience and its own polemical purpose within the larger framework of what was being written by other people at the time. If Russell was writing primarily to persuade, his work from the interwar period then needs to be placed within the rhetorical context of the public conversation of the time in order to understand more clearly the opinions he so effectively preached.[13]

Russell himself acknowledged the elements of performance in his ethical writings when he commented on his stance as a preacher in his reply to various critics in the anthology edited by Paul Schilpp.[14] He noted that "persuasion in ethical questions is necessarily different from persuasion in scientific matters," because ethics concerns the expression of desires rather than scientific truth, requiring the writer on ethics to "try and rouse these desires in other people."[15] He then went on to make explicit his own position as a preacher:

> This is the purpose of preaching, and it was my purpose in the various books in which I have expressed ethical opinions. The art of presenting one's desires persuasively is totally different from that of logical demonstration, but it is equally legitimate.[16]

Thus Russell could (and did) assail various people for their philosophical speculations about science and religion, for example, while at the same time utilizing a similar approach in his own sermons about science and ethics in the modern world. The difference, he would have said, was that where these people actually thought what they said was true, Russell himself merely wanted others to agree with an assessment that he preferred.

If, however, "rhetorical performance" was as important as I suggest, then the nature of the audience to which Russell played needs to be clarified.[17] The number of volumes on science, religion, and social ethics, and the multiple editions in which they appeared, is an indication of the size of the audience that wanted access to the debate in the interwar period. Similarly, the number of articles in popular periodicals indicates an equivalent interest among their readership. The proportion of books that appeared in inexpensive formats was substantial enough to indicate widespread interest.[18] The *Today and Tomorrow* series, for example, was priced at twenty-five cents in the United States and at one shilling in Great Britain. Similarly, essays were published in a less expensive pamphlet form; a number of Russell's essays first appeared in this format, such as the provocative "Has Religion Made Useful Contributions to Civilization?" published by E. Haldemann-Julius.[19]

I suggest that the audience for these debates was intelligent and literate, if not formally educated, and cut across the lines dividing social and economic classes. The cost of the publications was not prohibitive, and the existence of libraries, even in small communities, would make them available to those who had neither the means nor the opportunity to purchase them. Such an audience would be comprised of people who also would attend public lectures (lectures that often were subsequently published and therefore reached a far larger audience) or listen to such lectures on the radio.[20] A letter to Russell from Nathaniel Zalowitz, the editor of the English section of *The Jewish Daily Forward,* dated May 13, 1926, confirms these conclusions.[21] Russell wrote a total of fifty-three articles for the *Forward,* from May 30, 1926, up until his review of James Jean's *The Mysterious Universe,* published on December 28, 1930,[22] in a style similar to that of his other ethical writings from this period.

While I maintain later that Russell's conception of science constitutes a link between publications on otherwise seemingly disparate topics during the interwar period, I also believe that the inherent duality of his view of science as knowledge and science as technique is ultimately never resolved. Even later in life, he refused to connect his epistemology to his work on social ethics. In his response to Eduard Lindeman's essay on his social philosophy, Russell said: "I note with pleasure that he sees no necessary connection between my views on social questions and my views on logic and epistemology. I have always maintained that there was no logical connection."[23] However true Russell felt this statement to be at the time, it is misleading. To say there is no *logical* connection is not the same as to say there is *no* connection.[24] Lindeman added another important element to this discussion, which occasioned no objection from Russell, when after asserting "I see no necessary relation between Russell's epistemology or his metaphysics and his social philosophy," he amended his assertion by saying "the one element of inter-connection is his conception of the nature of science."[25]

Therefore if we consider that Russell's "conception of the nature of science" linked together works on epistemology, social ethics, and the relationship between religion and science, then I suggest there are grounds for claiming the inherent continuity of what Russell wrote during the interwar period.

To argue for continuity, however, is not to argue for either consistency or congruence. A creative tension between disparate ideas in his work has been recognized both by Russell himself and by some scholars. There is no need to attempt a grand synthesis of Russell's ideas, or to conclude, in the absence of such a synthesis, that his ideas are somehow inconsistent. Louis Greenspan, for example, reflected on the relationship between the crucial elements of science and liberty in Russell's work, much of which was written in the interwar period.[26] Calling Russell's views on the future of scientific organization and of individual freedom "incompatible prophesies," he pointed out how liberal ideals about individual freedoms conflicted with the degree of social organization made necessary by industrial civilization. The result, for Russell, was an ambivalence about the social virtues of science and the prospects of life in a scientific society. The conflict between "freedom and organization," Greenspan concluded, was one that Russell ultimately was not able to resolve.

Russell's emphasis on the dichotomy between his epistemology and ethics indicates at least his own inability to achieve a satisfactory synthesis of the two. Before the War, Russell had envisioned writing (in what Nicholas Griffin has called the "Tiergarten programme"[27]) two series of books, one on philosophical questions that would become more and more practical and one on social questions that would become more and more philosophical, until they coalesced somewhere in between. Although in the end the full Tiergarten programme did not materialize, Russell's work after the War maintained this kind of bifocal character in terms of his concern with the moral dilemma presented by "the old savage in the new civilization" and the existential dilemma posed by the metaphysical implications of the discoveries of modern science. I contend that his conception of science, considered as knowledge and as technique, provides the frame through which these two dilemmas were viewed by Russell.

Before proceeding any further, some clarification of terminology is necessary. I have opted to use "modern" in the same sense as I believe Russell and his contemporaries were wont to use it, both as successor to the values and ideas of Anglo-American society before the Great War and as harbinger of the many changes resulting from science and technology whose outlines they tried to discern. No further philosophical significance, therefore, should be attached to its use. The phrase "popular literature" is meant to indicate literature written for, and accessible to, the general public; most, if not all, of what Russell wrote in the interwar period may therefore be characterized as "popular literature." The terms "industrial civilization," "industrial society," "scientific society," and "machine civilization" are used interchangeably here, just as they were during the interwar period. Each was an attempt to express the character of the society that seemed likely to emerge from the rubble of the Great War, one whose future was to be shaped and controlled by science and industry. The word "technique" is meant below to include all of the possible applications of modern scientific knowledge. While Lewis Mumford in *Technics and Civilization* (1934)[28] attempts a more specific, historical interpretation of the term, Russell and his contemporaries, for the most part, avoid such specificity or philosophical precision in their use of it.

As for the terms *science* and *religion,* I intend to make some contribution toward the understanding of their usage in the literature of the interwar period. While more specific illustrations of this point will be made later, their usage suggests a definition of each term that was extremely elastic. Even Russell himself was prey to what could be viewed as an inconsistency of the times, making problematic an inclusive definition of his own perception of either "science" or "religion" in the interwar period. What precisely was meant by "metaphysics" in the interwar period I find equally unclear. For my purposes, I have regarded as "metaphysical" those attempts to formulate theories of reality that go beyond the phenomenal and descriptive limits of scientific knowledge. The line between "metaphysical beliefs" and "religious beliefs" is further blurred by those who considered their metaphysics "religious" in nature.

Finally, I offer the observation that because Russell was already a mature thinker before the end of the Great War, there are certain ideas and conclusions in his work during the interwar period that previously had found some kind of public expression. Similarly, his work did not end in 1938, and so his subsequent reflections on science and society (especially in the 1950s) might help explicate what he wrote in this earlier period. I maintain, however, what Russell wrote between 1919 and 1938 reflects emphases and concerns specific to the interwar period, requiring these works first to be interpreted and understood in that context.

NARRATIVE TABLE OF CONTENTS

Russell's conception of science is inextricably linked in his published works during the interwar period to two significant themes found in popular Anglo-American literature of the time as he grappled with the character of life in the Machine Age and the place of science and its applications in the society that was emerging from the rubble of the Great War. The first theme relates directly to the effects of the Great War, which demonstrated that moral development had not kept pace with technological progress, thereby putting increasingly dangerous weapons in the hands of the same old savage who had already shown himself incapable of handling them. This was popularly referred to as the problem of "the old savage in the new civilization." Further, while the applications of science were inescapably tied to the moral dilemma facing post-War society, science itself also generated considerable discussion on the meaning and purpose of existence. The second theme relates to the extensive public debate on the relationship between science and religion during the interwar period as a result of "the philosophical implications of the new physics." Recent discoveries in physics, stemming from Albert Einstein's work on relativity theory, called into question many fundamental assumptions about reality. As doubt was cast on existing scientific knowledge about the physical world, prominent scientists and clergy alike began to speculate on metaphysical topics in a manner that implied a convergence between the realms of science and religion.

In response to the problem of "the old savage in the new civilization," Russell opposed the representation of science as technique alone, calling for a moral outlook in science that incorporated more than merely an instrumental and utilitarian understanding of its applications. In response to "the philosophical implications of the new physics," he opposed the reconciliation of science and religion that others thought the new physics made possible. To Russell, "science" yielded much more than the technological applications that were shaping life in the Machine Age; it also yielded much less than the metaphysical truths that some were deriving from the recent discoveries of Albert Einstein and others. Yet it was "science," however construed, that constituted the foundation for industrial civilization, and would determine both its future and its meaning. Russell shared this conviction with those of his Anglo-American contemporaries who wrote hundreds of books and articles on these two themes during the interwar period.

That there were also continental influences, or individuals elsewhere who made some contribution to this discussion, or were in turn influenced by it, I have no doubt. Yet it is also clear that, to Russell and his contemporaries, their primary focus during the interwar period was on scholarship and popular literature in Great Britain and the United States. They expressed great concern about whether the benefits of science and technology outweighed the risks that they entailed, given the likely prospect of another world war. A new social ethic appropriate to the scientific society was therefore necessary if humanity was to survive its potential for self-destruction in the modern age. A survey of periodicals and books from this period yields the inescapable conclusion that the future of Western civilization was the subject of intense general interest and popular debate. The social effects of progress in science, and the urgent need to direct the changes that science generated, provoked a wide range of answers from Anglo-American authors.

Chapter 1 outlines the public conversation on the place of science in modern society and the potential threat posed by its applications by drawing on the work of representative Anglo-American authors who published articles and books in the interwar period.

The dilemma posed in terms of "the old savage in the new civilization" was a recurrent theme in Russell's published work during the period 1919–1938. He was not optimistic about the prospects of the society spawned by industrialism. At best, his assessment about whether the old savage could survive in the new civilization was ambivalent. At the heart of that ambivalence lay the problematic nature of modern science and how it might be directed for the benefit, not to the detriment, of the individuals who collectively had to live in the world it was in the process of creating. He thought it *possible* to improve the character and condition of humanity as well as society, but did he not consider such improvement inevitable, or even likely.

Throughout Russell's work in the interwar period there was a pragmatic awareness of the need to ensure the survival of civilization in an age where the means of its self-destruction were increasingly pervasive, and apparently uncontrollable. The Great War had made inescapable the conclusion that rationality, by itself, was not an adequate measure of human capacity, nor could it predict the actions that people might take. Emotion and impulse seemed to have played at least as great a role in the origins and conduct of the War as reason or logic. Russell understood that instinct, impulse, and emotion—as well as religion and morality—were also motivating factors in human behavior and thus affected the choices that humanity would make either to avoid the Next War or to hasten its arrival. He realized, as a result, that the future of Western civilization depended not simply on implementing the logical economic or political theory but rather on the more problematic task of developing a general philosophical outlook that was consonant with survival in an age dominated by science and technology.

Repelled by the vision of a mechanistic society in which efficient organization superseded individual freedoms, Russell struggled to articulate his understanding of "the scientific outlook" in which science was understood in terms other than merely its application to the problems of industrial society. Such an outlook, by nature, was philosophically imprecise. It might involve what technology made it possible to accomplish, or what science made it possible to understand, or the meaning that religious beliefs could ascribe to existence. Yet throughout the popular Anglo-American literature of the interwar period, there was the persistent belief that the modern age, or the Machine Age, was characterized by a changed perspective on the various aspects of social and intellectual life. Russell, in his turn, struggled to articulate a moral outlook different from the mechanistic, utilitarian, and instrumentalist one that would lead to the scientific society outlined in Aldous Huxley's *Brave New World*.

In Chapter 2, Russell's understanding of industrialism is elaborated, focusing on how he considered the mechanistic outlook more of a threat to industrial civilization than economic nationalism and on his efforts in *The Scientific Outlook* to articulate an alternative "moral outlook" that was neither utilitarian nor instrumentalist in nature.

In rejecting the mechanistic outlook for a "moral outlook" in science, Russell found himself in the company of those of his contemporaries who coupled the ethical question, "What must we do to survive in the modern age of science?" with the metaphysical question, "What does science tell us about the meaning of the universe?" Russell could not agree with Albert Schweitzer, however, that ethics are the product of worldview, and that a change in worldview must therefore precede a change in ethics. He also could not accept a positive role for religion in the formulation of a new social ethic for the Machine Age, objecting both to any social morality expressed in terms of religious creeds and to any reconciliation between the superstitious truths dogmatically asserted by religion and the progressive Truth discovered incrementally through the progress of modern science.

Forced by the implications of scientific technique for industrial society to search for pragmatic answers to the dilemma that technological progress presented, Russell considered religion during the interwar period almost exclusively in terms of its social and institutional character. Rejecting the idea that religion could be a source of knowledge or meaning outside of that provided by science, Russell excluded a religious or metaphysical basis for the social ethic of a scientific society, and seemingly for its "outlook" and "values" as well. A social ethic for Russell was a collective ethic, one to which people with varying personal convictions could still give assent, regardless of their individual beliefs about metaphysics or religious meaning. What individuals believed was immaterial; it was their collective response to moral questions that would define the ethic necessary to keep the old savage alive in the new civilization. The source of that collective response was problematic, provoking debate over whether there was any basis in reality for distinguishing between "good" and "bad," or whether (as he preferred) such qualities were merely an expression, writ large, of individual desires. Lacking a metaphysical component, however, such an ethic tended toward the very mechanistic and utilitarian outlook Russell wanted to avoid.

In Chapter 3, Russell's perception of religion during the interwar period is explored, illustrating his rejection of a positive social function for religion, especially in the formation of a social ethic appropriate to society in the Machine Age.

The effects of the new discoveries in physics on a materialistic view of the cosmos reverberated in religious and philosophical circles during the interwar period, suggesting the possibility there were both qualitative and quantitative dimensions to reality that were equally accessible to human inquiry and experience. As scientific knowledge about the physical world proved more tenuous in character than previously had been thought, and as claims about absolute scientific truths yielded to talk of mere probabilities, prominent scientists and clergy alike began to speculate on metaphysical topics in a way that implied a convergence between the realms of science and religion. They questioned whether there were sources of knowledge and truth other than the physical sciences, and other dimensions to reality beyond those that were material and quantifiable, and they wondered what all of these new ideas entailed for the meaning of existence and of the universe itself.

Russell himself was actively involved in disseminating at a popular level the discoveries of the new physics and their significance in understanding the nature of mind and of matter. His *The ABC of Atoms, The ABC of Relativity, The Analysis of Mind,* and *The Analysis of Matter* gained him a reputation and a large, popular audience. While he was very interested in the new possibilities of meaning that the physical sciences seemed to be providing, he was not sanguine about some of the philosophical or religious conclusions various influential people had reached. Thus when prominent scientists wrote popular books to advance their own metaphysical speculations, Russell felt compelled to repudiate at a popular level what others considered "the philosophical implications of the new physics."

In Chapter 4, the philosophical implications that Russell drew out of discoveries in the new physics are set out, along with the metaphysical implications expounded in the popular press by physicists Arthur Eddington and James Jeans and Russell's highly critical and equally public rejoinder.

On his part, Russell was convinced that science and religion were ultimately irreconcilable. Religion or metaphysics of any sort could not yield "scientific" knowledge of reality, and therefore could not lead to the discovery of truth about the universe. Without a basis in knowledge and truth, assertions about the meaning of life, beyond the level of personal conviction, were ultimately meaningless for him. Russell's support of the inevitable conflict between science and religion, which he located in the incompatibility between the provisional character of the truth associated with scientific knowledge and the dogmatic assertions of ultimate truth associated with religious creeds, doomed his attempt to distinguish science as the pursuit of knowledge from science as the application of technique.

These conclusions, reflected in *Religion and Science* (1935), came out of the context focused by two popular and public collections of ideas from his contemporaries. *Science, Religion, and Reality* (1925), edited by Joseph Needham, brought together diverse academic perspectives to explicate the essential dualism of science and religion within the new picture of reality provided by science, especially physics. In 1931, an entirely different tone was achieved through the publication of *Science and Religion,* a collection of short essays reworked from a radio series produced by the British Broadcasting Corporation, one that tended toward a genial and positive depiction of the need for religion in an age of science. *Religion and Science* was written as a response to the BBC series (though it included reprints of various earlier blasts in the popular press) and as Russell's attempt to deflect the public from concluding that there was some essential harmony between scientific truth and the creedal character of institutional religion.

In Chapter 5, some of the theoretical considerations involved in science and religion discourse are elaborated, including the nineteenth-century American origins of their supposed conflict in the work of J. W. Draper and A. D. White. The public conversation on the subject during the interwar period is outlined in terms of the contributions to the two collections of essays noted above. Russell's position is developed out of *Religion and Science,* including his defense of the conflict thesis and his rejection of a positive social role for mysticism or notions of cosmic purpose.

Russell's attempt during the interwar period to articulate a moral outlook in science and to solve, at the level of individual moral progress, the dilemma posed by "the old savage in the new civilization" was thus ultimately unsuccessful during the interwar period. In *Power: A New Social Analysis* (1938), Russell moved away from any possibility of locating the scientific outlook outside of the functions of human society. Any ethical impulse in society therefore had to be framed in the language and practice of power, making the morality of the individual less significant than the collective morality of the group. Without at least an "operational metaphysic" to which a group of individuals could assent, Russell found himself with no way of making the leap from personal to social morality that a modern social ethic required. Individual moral progress might be accomplished through education or whatever means, but there was no assurance it could ever be translated into the political or social structures of the new civilization.

In Chapter 6, I conclude Russell, by 1938, had invalidated the association of science and metaphysics in a way that ultimately left him no other grounds than the utilitarian exercise of power on which to establish a modern social ethic. These were the very grounds that he had earlier wanted to reject as part of his concern over the preservation of individual freedoms in a scientific society.

Thus the stage was set for the war that began in 1939. It is only after its dust begins to settle that Russell returns in the 1950s with renewed concern to the same issues in social ethics, human knowledge, the nature of science, and the exercise of political power that might still prevent the Next War—a war that in certainty would end not only human civilization but all life on earth.

Chapter 1

Science and the New Civilization

While fault could be found with individuals and political institutions for bringing about the 1914–1918 war and its aftermath, blame for the magnitude of the disaster was laid at the door of the science and technology that made such widespread destruction possible. The machinery that had revolutionized industry in the past half-century had also revolutionized the business of warfare, killing people in as efficient a fashion as it manufactured the commodities of an industrial society. The machine was no longer just seen by its critics as ugly or distasteful; it had demonstrated itself to be deadly as well. Scientific discoveries and technological developments therefore held at least as much threat as promise to people who had experienced the effects of machine guns, tanks, and combat aircraft.

The consequences of the Great War were thus tied to the prospects of the Next War by the implications of what modern man could do to himself in this new civilization, due to science and technology. Science was viewed with increasing suspicion in the interwar period, as the social costs of its achievements became apparent. When E. A. Burroughs, the Bishop of Ripon (in a remark he later claimed was offhand[1]), suggested at the meeting of the British Association for the Advancement of Science at Leeds in 1927 that humanity would benefit from a ten-year moratorium on scientific research, he provided an instant focus for the growing public unease over the social effects of technological change. Out of the carnage of the first modern war came the realization that human morality and ethics had not kept pace with technological development. The bishop's remark crystallized the debate about whether humanity could survive its technological potential for self-destruction in what came to be called the Machine Age.

There could be no doubt that moral development had failed to keep pace with technological progress, thereby placing ever-more dangerous weapons in the hands of the same old savage who, in the Great War, had already shown himself incapable of rational self-control. Science, in the form of technique, had thus created a dilemma in the aftermath of the Great War. Without some collective moral development that would enable people to control the new tools that science and technology provided, the catastrophic effects of their misuse threatened the continued existence of civilization. The experience of the Great War, however, inspired no great confidence during the interwar period that such moral development would happen quickly enough to prevent the inevitable catastrophe, thus rendering uncertain the future of Western society. This dilemma was depicted in popular literature as the problem of "the old savage in the new civilization," as Raymond Fosdick entitled a collection of addresses first published in 1928.[2]

Fosdick's metaphor of "the old savage in the new civilization" appears to have been taken from *Tantalus, or The Future of Man* (1924)[3] by F. C. S. Schiller, one of the *Today and Tomorrow* series of popular, mass-market publications in which various authors tried their hands at prognostications. Within five years of the appearance of its first volumes in 1924, the *Today and Tomorrow* series numbered more than 100 titles, providing a montage of the issues and opinions that preoccupied the Anglo-American public in the late 1920s.[4] In it, the problem of the "old savage" was identified with the human propensity for violence and destruction, a theme all too familiar in the period after the Great War. Merely developing science further in directions that might lead to improvements in weaponry, without either providing some external controls or ensuring individual restraint, seemed to make the Next War an unavoidable event.[5]

As a barometer of public opinion in the interwar period, *The Old Savage in the New Civilization* is an invaluable text, not because of Fosdick's literary brilliance or originality but because he represented at a popular level the ideas about science, technology, and the future of industrial society that had made the transition from the university to the corner store. Perhaps more so than at any other time in this century, the effects of science and technology were immediate and obvious in the lives of ordinary people, and the debate over the consequences of industrial development was carried on in ways accessible to the general Anglo-American population. As a piece of popular literature, *The Old Savage in the New Civilization* also provides—in its title—a convenient metaphor by which to depict the context of the period in which Russell and his contemporaries wrote.

While the effects of science and its applications in technology were to be found everywhere, the opinions of experts about what the future might hold often were contradictory, serving only to increase public anxiety regarding what the latest round of inventions or scientific discoveries might mean. Sir Frederick Soddy, the British chemist and author of popular books on science, had warned by 1925 of the terrible consequences of splitting the atom if this knowledge was applied to making weaponry.[6] Yet, at the same time, J. B. S. Haldane, the British biologist, and Robert A. Millikan, the second American ever to win the Nobel Prize in physics (in 1923), were in agreement that atomic energy was irrelevant to the future of civilization.[7] Similar disparities of viewpoint flourished in virtually every field of science and technology. Not until after the bombing of Guernica by the Condor Legion during the Spanish Civil War, for example, was the role of airpower in future warfare finally accepted, despite dozens of earlier demonstrations and predictions.[8] What seemed apparent, however, to all but its mostly blindly enthusiastic promoters was that the practical benefits of science for society were disturbingly equivocal.

From the soapboxes of the periodical press to the lecterns of famous universities, the depiction of future Western society in all of its aspects both caught and fueled the popular imagination during the interwar period. Science was central to these depictions, both for the benefits it conferred and for the potential dangers scientific development entailed. What concerned many people was the relationship between science and technology, between the knowledge attributable to science and the applications of that knowledge in the form of industrial technique.

"Science is the engine of industrialism, and industrialism rules the modern world," C. E. Ayres wrote in 1931, for example, "but what that process (of industrialization) means and is going to mean for civilization and just how the fate of science is linked to the fate of industry no one knows."[9] Calling science "a peculiarly insidious achievement and machine industry a strangely equivocal triumph," Ayres complained that the old social problems had not been solved, but industrialization seemed even to have multiplied them.[10] Depicting science as "the combined techniques of numbers and machinery," he observed that the difference between pure and applied science was only one of degree: "Science is the theory of industrial technology, and industrial technology is the application of science. Neither is historically possible nor logically intelligible except in terms of the other."[11] The irony of science, Ayres said, was that while it was expected to provide the answers to the modern social problem, science had produced the problem in the first place.[12]

For many people, it seemed as though science was responsible for creating the new Machine Age into which humanity had been propelled. In fact, the category of "machine age" does not appear in the *Readers' Guide to Periodical Literature* until the 1929–1932 volume, which lists forty-seven different articles, most with "machine age" in the title. Yet this attitude disregarded the obvious fact that "technology" was as old as the human race, and that "machines" had a history almost as long. Lewis Mumford's landmark book *Technics and Civilization* (1934) may be considered the first modern attempt at exploring the relationship between technology and society, and it was devoted to precisely this point.[13] Similarly, in *Men and Machines,* Stuart Chase, a prominent American author who explored the problems of the Machine Age, tried to come to a more balanced perspective on the relationship between humanity and its machines in the new civilization.[14] He evaluated the progress that had been made in controlling the external world through machines and laid out the equivocal nature of this triumph in the form of a balance sheet. Yet, in his analysis, he also detailed the way in which Western society had been shaped and directed by a host of machines since its earliest times, and he noted that much of the machinery presently in use had been around for the better part of a century. Chase observed that it was not so much the machines that were giving Western society trouble as its mores, which he said "are in flux and conflict, while many of its institutions and most of its commodities are becoming ever more complex."[15]

Whatever the long history of technology, however, something had changed in the post–War world, and it was not simply a matter of popular perception. Chase observed in passing that it might be better to call the present time the "Power Age," as Henry Ford had dubbed it, instead of the Machine Age,[16] and this was a perceptive comment. Whether the machines had existed before or not, the stakes for their use had been revealed in the Great War and its aftermath. Speed and power, change and innovation were the hallmarks of industry in the interwar period. Not only had society been unable to control the effects of its machines in the Great War, it was becoming increasingly problematic whether such control would ever again be possible. While the phrase "autonomous technology" has been coined more recently, the sentiment was reflected in the anxious words of people after the Great War who felt the future held at least as much threat as promise.

The discussion about the future of Western civilization itself, however, was largely an Anglo-American one during the interwar period. Though the Next War was indisputably tied to the probability of ideological as well as economic conflict, the absence of an immediate revolutionary threat in Great Britain or the United States allowed Anglo-American thinkers to view the prospects of *their* modern society in more dispassionate terms. In *Brave New World,* Aldous Huxley was concerned with how industrial totalitarianism could lead to political tyranny, and not the reverse.[17] Similarly, the idea of technocracy fomented political debate within the United States.[18] Thus the "new civilization" tended to be an extension of their own, recast in the light of modern science and technology, to be sure, but without considering the radical discontinuities made likely by a political revolution.

One interesting feature of the Anglo-American debate on machine civilization is that while its critics tended to be British, its promoters tended to be American. In *Brave New World,* for example, the dates are determined before and after Henry Ford, and the application of "Fordism"[19] to the production and education of babies was intended to evoke disgust, not admiration, in the reader. Of course, the same things that were transforming life in the United States were also transforming life in Europe, but due to the devastation of the Great War, and perhaps to a more conservative attitude toward social change, these transformations were less immediately dramatic.

J. F. C. Fuller's *Atlantis: America and the Future* is a vivid illustration of the dangers that were envisioned in the uncontrolled development of the machine civilization that could most easily be seen in the United States.[20] Saying that "American civilization is European civilization minus some two thousand years,"[21] he fumed: "Spiritually the country is a corpse, physically a terrific machine. Materialism is the tyrant which rules from ocean to ocean, and its backwash is superstitition and an effervescing froth of cranks."[22] Calling Henry Ford "only a rather pronounced product of his age," he complained that "not only is the Ford car produced in its millions but also the Ford mind."[23] Herein lay the real danger of Fordism: "In the Ford system, the worker ceases to be a human instrument, in place he becomes a mechanical tool; craft disappears, and the man is moulded by the machine."[24] Fuller's concern was heightened by his perception that in the hustle of American life "there is little or no time for reflection, yet it is reflection which begets forethought, whereby success is foreseen and disaster forestalled."[25]

Michael S. Pupin, a famous American electrical engineer, responded to various European criticisms of American machine civilization in his *Romance of the Machine*.[26] "Some of [these Europeans] write of it as an American plague," he said, "which is invading Europe and threatening to undermine its ancient culture. 'Americanism' is the name which these writers frequently give to this new civilization, and Americanism, is their opinion, is sordid materialism."[27] He then went on to examine the European indictment against "American science and engineering, which they hold responsible for the evils of the so-called Machine Civilization."[28] Pupin extended the idea of the machine to represent all living things, making it not a product of human manufacture but an inherent aspect of nature, which humanity of late has begun to emulate in its own inventions. As for the charge of worshipping the "heathen idol" of the machine, Pupin said that the worship of scientists and engineers was directed to the scientific principles which lie behind their machines, hailing "Galileo, Newton [and] Faraday" for having laid the foundation of machine civilization.[29] In what rapidly degenerates into a panegyric to American ingenuity in applied science, he extolled the "spirit" inherent in every machine, "which keeps alive our 'passion for profound knowledge,'" and he solemnly intoned, "He who does not feel this spirit cannot understand the story which these machines are telling."[30]

LIFE IN THE MACHINE AGE

The debate over whether science was the author of humanity's fortune or its misfortune was subsumed within a larger discussion about the character of life in the Machine Age. The extent and rapidity of the changes taking place in Western society were undeniable. Science was fashioning a new machine civilization, quite apart from providing the old savage with more efficient tools of destruction. Machines altered work patterns, living arrangements, transportation, recreation, and leisure, and through the media of radio and the press, political and social thought as well. Life was different in the Machine Age, and the nature of that difference was cause for both celebration and concern.

Four representative examples of the extensive, popular Anglo-American literature illustrate the concerns of Russell's contemporaries about the relationship of science to the new civilization. First, Raymond Fosdick's analysis from *The Old Savage in the New Civilization* is presented. This is followed by what Robert Millikan, Nobel laureate in physics and a frequent contributor to periodicals in the 1920s, had to say in response to Fosdick and to the Bishop of Ripon's call for a moratorium on scientific research. William McDougall's pungent criticism of "the responsibility of science" for the problems of the Machine Age leads into J. W. N. Sullivan's discussion of the misplaced "tyranny of science" in popular culture.

There are certain elements common to the work of these four authors. Science is integral to their depictions of present and future Western society. They are aware of the inevitable dependence upon machinery in the modern age and the interdependence of the various parts of industrial society. They recognize the dangers inherent in the uncontrolled development of technology and look to an understanding of science-as-knowledge for the means to acquiring such control over science-as-technique. Values are understood to be an important component of the popular perception of science, though the source of these values is unclear. Finally, the survival of the old savage in the new civilization is seen to be more dependent upon the education of individuals into a new outlook, than upon solutions to the political and economic problems of the post–War world.

The Mechanical Circle

Raymond Fosdick asked pointed questions about the future of humanity in the Machine Age into which it had been propelled by scientific progress. All of his addresses in *The Old Savage and the New Civilization* were tied together by "a single theme," which he identified as "the new civilization into which modern machinery has plunged us, and the struggle of mankind to keep abreast of it."[31] The "supreme question" before his generation was this:

> Humanity stands today in a position of unique peril. An unanswered question is written across the future: Is man to be the master of the civilization he has created, or is he to be its victim? Can he control the forces which he himself has let loose? Will this intricate machinery which he has built up and this vast body of knowledge which he has appropriated be the servant of the race, or will it be a Frankenstein monster that will slay its own maker? In brief, has man the capacity to keep up with his own machines?[32]

The problem, as Fosdick saw it, was that science would not wait for humanity to develop a morality adequate to cope with the new choices presented by machine civilization.[33] Like Russell, Fosdick questioned whether the old savage could be improved in time to prevent his own destruction at the hands of the technology he had created.[34]

Through his own work and his writing in the popular press, Fosdick had a substantial influence on attempts to keep the old savage alive in the new civilization. He was appointed Under-Secretary General of the League of Nations in 1919 for one year, and he was one of the prominent figures in the unsuccessful campaign to get the United States to join the League.[35] He also urged the development of "the sciences that relate to man" in order to cope with ongoing scientific progress.[36] Succeeding his brother, Harry Emerson Fosdick, as a trustee of the Rockefeller Foundation in 1922, Raymond went on to become its president from 1936 to 1948.[37] During his trusteeship, the foundation shifted its focus for research funding away from the so-called "hard sciences" of physics and chemistry (responsible for the weapons that the old savage was unable to control) toward the "human sciences" of psychology, sociology, anthropology and biology.[38]

Despite these sentiments, however, it was obvious to Fosdick that the advent of industrialization meant machines had become indispensable to the future of civilization, regardless of the social cost they extorted. The problem with an opposition between the individual and the machine was that society now *needed* machines. Fosdick's depiction of this fact was to the point: "Stop the machines and half the people in the world would perish in a month." It was no longer a question of whether the machines were needed, but of how they might be managed to create a future in which people would want to live. It was not a question that could easily be answered. "We know now that we are not completely the masters of the machines we have created," Fosdick said. "Their pulsations we can control, but their consequences control us."[39]

For Fosdick, the "mechanical circle" of modern industry led to standardization and uniformity in society: "Idle machines mean starvation to the millions of people whom they have brought into the world; active machines mean a surplus of goods beyond the immediate capacity of the race to consume."[40] The problem then became how to encourage consumption, "not how to produce goods but how to produce customers; not how to develop output but how to intensify consumption."[41] People have to be persuaded to want what they have not wanted before, he said, and this required that goods be inexpensive; for goods to be inexpensive, they must be manufactured in quantity.[42] To manufacture goods in quantity, their production must be standardized, and Fosdick saw this impulse toward standardization everywhere: "Standardization is in the air," he said. "It even extends to standardized divorce laws and standardized building and plumbing codes."[43]

The interdependence of society in the Machine Age was in part a consequence of the same forces that promoted standardization. The supply of raw materials for the machines, as well as the need for markets for what they produced, was seen as a global enterprise. Should any portion of the global economy be disrupted, its effects would be felt in many other sectors. This made the problem of controlling the global economy even more difficult, for interdependence increased complexity in the system and made it more difficult to manipulate, while the consequences of any individual action became far-reaching.

As an illustration of such far-reaching effects, Fosdick enumerated the ways in which American consumer culture was already spanning the globe in 1928. "All around the world," he complained, "the habits and possessions of men are shaking down to fixed, common levels." The process was "even more pronounced" in the United States, where "the material side of life . . . is fast developing a sameness, a uniformity, a monotony without parallel in history over so wide a geographical area."[44] This uniformity was one of the unforeseen consequences of life in the Machine Age, and Fosdick blamed, in particular, "quantity production, advertising, and the new methods of communication and transportation" for "breaking down the differences which hitherto have made of civilization a garment of many colours."[45]

The drab picture Fosdick painted of modern mechanical culture assumed a more ominous cast when he considered the effects of such uniformity on individuals. "Common physical surroundings and possessions seem invariably to foster common mental reactions," he warned, observing that "there is something about mass production and distribution of goods that suggests mass production and distribution of ideas."[46] He noted the "uniformity of taste and thought" across the country that was promoted by the press as a consequence of wire services and "syndicated opinions," and he warned that the development of the radio was likely to accentuate the pressure toward intellectual conformity, with "audiences of five and ten million people listening to the same voice" an "almost daily phenomena."[47] These tendencies would lead, he feared, to the "tyranny of the majority" and to the suppression of individual thought, a fear that was shared by a number of writers on the future of the machine civilization.

The irony, of course, was that for Fosdick and others, the only real hope for the future of the "old savage in the new civilization" lay with the individual citizen, whom the machine civilization tended to devalue. Calling the rule of the majority a "clumsy device," Fosdick said it was "nonsense" to credit it with "a kind of centralized infallibility and to proclaim that the voice of the people is the voice of God."[48] Instead, like Russell, he asserted there was "no social good apart from individual good,"[49] and he applauded the virtues of nonconformity.[50] He felt that this was particularly important in the Machine Age, "For science has armed majorities with instruments of persuasion and coercion more effective than any which they have previously wielded, and the individual must seek protection against the new usurpations of society."[51]

Acknowledging that some would say this was "dangerous talk," he called for "a skepticism of intellectual authority, a distaste for unruffled unanimities, a toleration of differences."[52] Only in this way would society continue to change its institutions as well as its technology, he said, charging that the real danger to civilization lay not in change and progress but with those who tried to restrict it and to maintain the status quo:

> The peril is that under pressure of intrenched and satisfied majorities we shall stone the prophets once too often. The danger is that we shall cling to the shell of our social and economic institutions too long after they have been outgrown, adhere to the husk and form of ideas too long after they are dead.[53]

Science and Modern Life

Despite describing Raymond Fosdick as "one of the best informed and most intelligent of living Americans," Nobel laureate physicist Robert A. Millikan attempted to refute Fosdick's pessimistic conclusions about the relationship between science and modern society.[54] To the blame Fosdick attributed to science for the devastation of the Great War, his pointed rejoinder was that war was nothing new, and that "every scientific advance finds ten times as many new, peaceful, constructive uses as it finds destructive ones."[55] Noting the beneficial social changes of the previous fifty years due to science and its applications, Millikan ridiculed the idea that either had been detrimental to Western civilization.[56] In response to the concerns regarding modern social problems, he blamed instead "the perpetual motion cranks" of literature and art who promoted individual license, rather than social responsibility, and he said that it was from them, not from science, that "the chief menaces to our civilization are now coming."[57]

In his own collection of addresses, entitled *Science in the New Civilization* (1930), Millikan also opposed the Bishop of Ripon's call for a moratorium on science, noting that the bishop later qualified his call by specifying the need for "a vacation for physics and chemistry and the parts of biology not associated with the improvement of health and the alleviation of suffering."[58] While Millikan agreed with the bishop's concern regarding the control of science, he did not agree with his response. He felt it impossible for physics and chemistry to "take a holiday without turning off the power on all the other sciences that depend on them," firstly, and secondly, that along with genetics, these were "the great constructive sciences which alone stand between mankind and its dire fate foreseen by Malthus."[59] In other words, science was a sufficiently interdependent enterprise that no single part could be shut down without affecting all of the other parts; and science, however responsible it might be for various social evils, was also the only bulwark between humanity and the disasters contained in a Malthusian future. While agreeing with the bishop that there was a need to distrust "the wisdom, and sometimes even the morality, of individual scientists, and individual humanists, too," Millikan said that a holiday for science was "both impossible and foolish."[60] Instead, what society needed was a reconstructed educational process that could produce "broader gauge and better educated scientists and humanists alike."[61] Regarding the dangers of entrusting new weaponry to the same old savage, Millikan asserted that modern science had made it impossible for nations to conduct themselves as they had before, on threat of mutual destruction, requiring them to find ways of settling conflicts that did not involve recourse to warfare. Rather than proposing a moratorium on research, he pointed to the rapidity of scientific change as evidence that a rational approach to education might change human nature quickly enough to help Western civilization survive.[62]

As much as he disputed the assessments of the Bishop of Ripon or Raymond Fosdick, however, Millikan also recognized the problems that confronted "the old savage in the new civilization." He attributed at least some of the blame for the current "craze for the new regardless of the true" to the rapidity of social change made possible by modern science and technology.[63] While defending science against the charge it made material values more important than spiritual ones,[64] he also admitted the need for values that would direct the future development of science and its applications on a constructive path. Life in the Machine Age had become different because of science and technology, and this difference was a cause for some concern to Millikan as well.

The Responsibility of Science

William McDougall's boldly titled book, *World Chaos: The Responsibility of Science* (1931), credited a general fixation with physical science for the dilemma confronting Western civilization in the interwar period:

> The distinctive feature of our civilization is our Science. The thesis of this little book is twofold: first, that physical science has been the principal agent in bringing about the very rapid changes in our social, economic, and political conditions which are the source of our present troubles; secondly, that in the development of the neglected social sciences lies our only hope of remedy for those troubles.[65]

McDougall reached many of the same conclusions as did Fosdick, though he delved more deeply into the causes of the problems with science that Fosdick was content only to enumerate. The physical sciences, he said, were responsible for creating a level of complexity in Western civilization "which far outruns our present understanding and control."[66] Accepting the premise that "the civilization of any people reflects the state of its knowledge, and is in a large measure determined by that knowledge," dissatisfaction with civilization should lead us to consider whether there was not "some radical defect in our knowledge, more especially in the systematically organized part of our knowledge which we call Science."[67] Noting that modern civilization, though not founded upon science, had been radically transformed by it, McDougall asserted that science was responsible for giving modern civilization its distinctive "quality": "It undergoes perpetual and rapid change, and . . . its ideal is progress rather than stability."[68] Despite this analysis, he rejected the Bishop of Ripon's call for a moratorium on science, for much the same reasons as Millikan, and he urged the moral development of those who would control science as the only way in which to direct the future course of industrial civilization in a positive direction.

To counter-balance a society rendered increasingly "top heavy" by industrial organization and its reliance upon machines, and "lopsided" by the applications of physical science to manufacturing alone, McDougall, like the trustees of the Rockefeller Foundation, thought more attention needed to be paid to the social and biological sciences. The "radical defect" in our knowledge, he believed, was the overwhelming reliance on the physical sciences which, though they could create the complexity of an industrial civilization, could not provide any means for its control.

To make matters worse, all of the traditional wisdom that had previously governed human relations was rendered "utterly inadequate" by the effects of physical science and its industrial applications upon society. "We are compelled to try to live by the light of Science," he said, "and alas! we have no Science to guide us."[69]

McDougall's complaint about the general fixation with the physical sciences had merit, for it was evident to all that the chief applications of "science" were to be found in relation to the control and manipulation of nature. While proponents of the so-called "social sciences" talked about directing the course of social development, and others spoke eloquently about the possibilities of biological manipulation to adapt human beings to an industrial society, these remained highly speculative in comparison to the rapid advances in the field of technology. In the interwar period, therefore, it is not surprising to find that no specific distinction was made between "science" and "technology." Science was held to mean what went on in the laboratory, and held responsible not only for whatever technological advances resulted but also for the uses to which they were put. Yet, as McDougall rightly observed, none of the technological advances that physical science bestowed on humanity came with the appropriate guidance or wisdom to direct its use. "Mere machines" seemed mere no longer; to many, the products of scientific technique appeared to have a life of their own and, like the story of Frankenstein's monster, this did not bode well for the new civilization.

The Tyranny of Science

In his contribution to the *Today and Tomorrow* series, *Gallio, or The Tyranny of Science* (1927), physicist J. W. N. Sullivan lampooned the popular misunderstandings of science that had recently emerged, in part as a consequence of the Great War, and in part as a result of the new discoveries in physics.[70] Science was credited with providing more knowledge of the universe than it actually did, said Sullivan, and the methods of the physical sciences were misapplied in other areas of "science" in a way that occasioned more absurdity than enlightenment.[71] Chief among the absurdities of the misapplication of science, he said, was the idea that it somehow "proved" the ultimate meaninglessness of life:

The notion that we live in a purposeless universe is so opposed to the mental habits we have inherited that it is a matter of the greatest difficulty to bear it constantly in mind. Most of the people who hold this belief today would not do so but for three reasons: the disillusionment caused by the War, their respect for science, and their belief that science preaches materialism.[72]

Sullivan did not address the disillusionment caused by the Great War, beyond the wry observation that while the experience was "consistent with the belief that man is a developing spirit . . . it is certainly a proof that he is not very far developed."[73] He focused on the other two problems: the popular belief in materialism, and the credulity with which science was generally regarded. Commenting that "the respect for science, is, I believe, on the whole rather overdone," he was unsparing in his criticism of what he would go on to call, in a subsequent book, "the limitations of science."[74]

In pointing out the gulf between popular and scientific perceptions of what the new physics implied about materialism, Sullivan showed exasperation with those who claimed a scientific basis for outdated ideas. For some reason, "At a time when the physicists are abandoning materialism, the artists are accepting it. They are accepting, as the last word of science, a picture of the world that belongs to the early bad manner of physics."[75] He blamed such popular support for materialism on the psychological effects of the Great War: "It is a curious but indisputable psychological fact . . . that the sight of a large number of naked human bodies makes it difficult to believe that they are animated by immortal spirits possessing an eternal destiny."[76] Thus, while the relationship between mind and matter was ostensibly a scientific question, he observed that its treatment in the popular arena involved more faulty philosophy than it did good physics.

Sullivan attributed conclusions about the lack of cosmic purpose to an interpretation of human experience rather than to what science revealed about nature:

> We have seen that the philosophy that regards man as a meaningless accident in an alien universe receives no support from modern physics. The true ground of that philosophy is now, as it has always been, the apparently meaningless misery that forms part of life.[77]

What modern physics revealed was a convergence between values and numbers, in which quantities had to be supplemented by qualities in order to accurately represent the nature of the universe. Sullivan rejected the idea that there was a conflict between science and mysticism, or art, because such a conflict depended upon "an old-fashioned conception of the status of physics" which was "not up-to-date."[78] The "tyranny of science," for Sullivan, consisted in the attempt to exclude qualities or values from what could be known about that reality:

> But if we do not adopt the materialist principle we may assert that moral and aesthetic values are as much a part of the real universe as anything else, and that the reason why science does not find it necessary to mention them is not because they are not there but because science is a game played according to certain rules, and those rules have excluded these values from the outset.[79]

Sullivan concluded that the "new outlook" would have to regard the world "as an evolutionary process, where 'patterns of value' emerge . . . an interconnected whole [in which] the separation of mind from matter, and mind from mind, will have to be replaced by a conception which regards these distinctions, in their present form, as unreal."[80]

The association between mind and matter, the demise of materialism, and the assertion of the reality of "values" in the universe had far-reaching implications for the relationship between science and religion in the interwar period. Sullivan was not alone in trying to identify the "new outlook" that was attempting to incorporate these changes. What science could tell us about the universe and the limits of knowledge derived from science were equally crucial components of this new outlook; so, too, was the question as to whether "values" had any place in the universe that science revealed.[81] The discussion of values, however, had more than a casual significance: the source of values, or of morals, and thus of a social ethic, was a crucial question for the generation that was trying to avoid the Next War. If values had an independent existence, if facts about morality could be known with as much certainty as facts about science, then perhaps some way could be found to create the global ethic that the old savage needed in order to survive. Sullivan was very much aware of these implications:

> On the new outlook, the function of the arts is to communicate knowledge and, moreover, the most valuable kind of knowledge. Art, much more than science, expresses the concrete facts of experience in their actuality . . . Not only art, but morals, acquire vastly greater importance on the new outlook. Morals is no longer a purely private concern, expressive of a particular human constitution in an alien, strictly non-moral universe.[82]

Sullivan did not himself conclude that humanity lived in a friendly, moral, and purposive universe, but there were others at the time who did.

Thus questions of what science revealed about the universe, or what implications science might have for the understanding of meaning and truth, were ultimately inseparable from the applications of science that were creating a civilization doomed by its inability to use such power wisely. After the Great War, it was generally believed there was some kind of new outlook that had come to characterize life in the Machine Age. Precisely what that new outlook was, and what it needed to become if the old savage was to survive in the new civilization, was a subject more important than any other to Bertrand Russell during the interwar period.

Chapter 2

The Individual and the Machine

If any one aspect of philosophy had preoccupied Russell before the Great War, it was the need to understand the axioms, or the initial premises, on which other knowledge was based. He wrote about the foundations of geometry, the principles of mathematics, and what it was possible to know about the physical world, becoming one of the founding figures in modern analytic philosophy as a result.[1] Because he saw the Great War as the result of a catastrophic failure in the outlook and principles of Western society, he was also not content to leave the philosophical basis of industrial civilization unchallenged.

The Great War made it clear that modern powers of destruction were not controllable by existing political structures, nor were they made unthinkable by moral strictures against the use of what science and technology provided. Even as Russell admitted the inescapable nature of industrial development in the modern world, he saw that the immediate danger lay in the continuation of antiquated nineteenth-century attitudes and institutions that were incapable of controlling such development. In *Freedom and Organization* (1934), he concluded that the nineteenth century had failed to create an international political organization to match the economic conditions created by science and industry, and so, "in a haphazard way, as a result of technique unguided by thought, it created economic organizations which its philosophy did not teach it to control."[2] In particular, he said, no one had appreciated "the part played by organization in a world ruled by scientific technique."[3]

For Russell, the initial premise of industrial civilization was science, "non-political itself, but controlling all political occurrences."[4] His search for the axioms at the heart of the Machine Age yielded his understanding of the duality of science-as-knowledge versus science-as-technique, and the centrality of power as the fundamental principle of the scientific society. Russell evaluated the political and economic conditions of post–War Western society, but even as he identified the problems associated with "economic nationalism" and the likelihood that the Next War would be a class or patriotic war, he realized that the root problem of the scientific society was its mechanistic, utilitarian outlook, which devalued the individual in the interests of social organization.

Prospects of Industrial Civilization (1923) was Russell's most significant attempt to analyse the plight of Western society and to suggest some possible avenues for avoiding the Next War.[5] In it, he left behind nineteenth-century political ideas in an attempt to articulate the political, economic, and scientific forces shaping society in the Machine Age. Recognizing that "a new economic mode of life brings with it new views of life, which must be analyzed and subdued," he had come to believe that "the important differences in the modern world are those which divide the nations living by industrialism from those which still live by the more primitive methods."[6] Thus, he said, "the important fact of the present time is not the struggle between capitalism and socialism but the struggle between industrial civilization and humanity."[7] Russell contended that the "sacrifice of the individual to the machine" was "the fundamental evil of the modern world,"[8] and so he struggled to find a way in which individuality might be conserved in the midst of the social organization required by the scientific society.

Russell saw further industrial development as inevitable, and insofar as it involved an increase in knowledge and economic prosperity, it was to be desired. Unfortunately, he noted that "machinery, which is physically capable of conferring great benefits upon mankind, is instead inflicting untold evil, of which the worst may be still to come."[9] The problem lay not with industrialism itself but with the continuation into the Machine Age of outdated ideas, which threatened to bring about the Next War before those benefits could be achieved. He identified the two culprits as private property and nationalism, and he called for them to be replaced by "some form of public ownership," or socialism, and by internationalism.[10] Without socialism and internationalism, Russell thought industrial civilization could be counted upon to destroy itself within the next 100 years.[11] There was a third culprit, however, in "the mechanistic outlook." It had to be replaced by a humanistic outlook, "which values mechanism for its extramechanical uses, but no longer worships it as a good in itself," if the society made possible by the machine was to be fit for individuals to live in.[12] Russell came to realize that, whatever the philosophical arguments about capitalism or socialism, without such a shift in the outlook of industrial civilization the scientific society would not be one in which he wanted to live. The mechanistic outlook was the product of science considered only in terms of its applications, not in terms of knowledge or truth, and this Russell found distasteful. The prospect of a regimented, organized society, in which the freedom of the individual was subordinated to the interests of the State, was therefore tied to the way in which science was viewed by those who wielded the power that technology bestowed.

THE NATURE OF INDUSTRIAL CIVILIZATION

Russell's first prognostication on Machine Age society in light of the Great War had been *Principles of Social Reconstruction*. He later claimed that it was both spontaneous and unlike anything he had previously written.[13] Since he characterized the contents of this book as "a philosophy of politics based upon the belief that impulse has more effect than conscious purpose in moulding men's lives,"[14] the actual spontaneity of his design in writing it may be viewed with some suspicion. It is best considered Russell's reaction to the conundrum posed by the rapid about-face in European society from pacifism to nationalism, from rationality to irrationality, which he witnessed at the outbreak of the war. As a rationalist, he could not comprehend the nationalistic fervor that swept up even his colleagues from Cambridge into the tragic binge of irrationality that was the Great War. Accordingly, if human society was at root as impulsive as people's reactions to the war made it appear, then any post–War society that hoped to evade a similar conflict in the future had to find a way of channeling those impulses into more creative, less destructive pursuits.

Principles of Social Reconstruction was written well before the worst battles of the Great War, before the Bolshevik Revolution, and before the entry of the United States into a conflict that had all but exhausted its European protagonists. It is therefore not surprising that the meditative style and contemplative tone of *Principles* is markedly different from Russell's post–War reflections on industrial society, highlighting the watershed effect that the Great War had on his intellectual concerns. Russell had no clear idea about what needed reconstructing, because much of the damage had yet to be done, and he admitted as much.[15] His proposals were more abstract than practical. The first "principle of reconstruction" he proposed was that, "the growth and vitality of individuals and communities is to be promoted as far as possible"; the second, that "the growth of one individual or one community is to be as little as possible at the expense of another."[16] To satisfy these principles, there needed to be an "integration, first of our individual lives, then of the life of the community and of the world, without sacrifice of individuality."[17]

If there was one lesson the Great War was to drive home in agonizing detail, however, it was that industrial civilization was built upon the sacrifice of individuality. Millions of lives had been fed indiscriminately into the machinery of modern warfare, and neither intellectual gifts nor social standing made a difference regarding who was slaughtered. Individual soldiers meant less to the war effort of a nation than an individual automobile meant to the Ford Motor Company. This sobering fact lay behind Russell's later contention in *Prospects,* that "the mechanistic outlook" was as much of a threat to the future of civilization as capitalism and nationalism were generally recognized to be. What most threatened the place of the individual in the post–War world, in other words, was "the mechanistic conception of society,"[18] not simply industrialism.

Russell understood "industrialism" to be "an extension of the habit of using tools," or "essentially production . . . by methods requiring fixed capital."[19] The purpose of "using tools" was to produce commodities that satisfy "our needs and desires."[20] He identified five qualities required by a community for industrialism to be practiced successfully: a large organization of workers devoted to a common task; willingness on the part of those in control of this labor to forego present satisfaction for the sake of greater wealth in the future; a sufficiently stable government to make that future wealth probable; a large body of skilled workers to accomplish the difficult tasks usually part of industrial production; and "a body of scientific knowledge, to make and utilize mechanical inventions."[21] Russell observed that industrialism, left to itself, would eventually organize the whole world into "one producing and consuming unit," for there were no inherent limits to the growth of large-scale organization.

Industrialism made society "more organic," in that all of its parts, like the parts of an organism, became interdependent for their continued survival:

> In an industrial community, no man is self-subsistent; each man takes part in a process which produces a great deal of some commodity, or of some machine for making commodities, but no man produces the variety of commodities necessary for preserving life.[22]

Everyone, from the laborer to the capitalist, was dependent upon the continued association with other members of an industrial community for their survival. Such interdependence, moreover, rendered an industrial society more vulnerable to disruption, for there were "vital organs" in the social body whose destruction or impairment affected every part of it to a greater extent than would be the case in a more primitive society. It was inevitable, therefore, that government would assume greater importance and that its control over the lives of individuals would be extended further as a society became more industrialized. Because the actions of individuals have a greater effect on the whole fabric of an industrial society, individual liberty had to be curtailed in the interests of the community. This was the root of the conflict between the individual and the machine.

While industrialism had made strides toward the social control of individuals, however, nothing substantive had been done about the international control of nation-states. To avoid the Next War, there needed to be an equivalent means of controlling individual states, in the interests of industrial civilization as a whole. Russell characterized nationalism as "a development of herd-instinct," and as "the habit of taking as one's herd the nation to which one belongs."[23] Calling rivalry "part of the instinctive apparatus of human nature," he said the essence of nationalism was the extension of that sense of rivalry to relations between nations. As long as the majority continued to feel that its only social obligation was to its nation—even at the expense of other nations—no diplomatic or political efforts alone could produce "a tolerable world."[24] The threat posed by more destructive weaponry, in the context of nationalism, actually increased the likelihood of war.[25] Unless the destructive effects of nationalism could somehow be mitigated, Russell warned, there seemed to be little hope for humanity, "except in a total collapse of the industrial system."[26]

Internationalism, to Russell, was "primarily a matter of world-government,"[27] some organization or state sufficiently powerful to impose its decisions on humanity and to regulate the relations between nations by law instead of having them resort to the kind of warfare that could destroy everyone. This would have the effect, then, of preventing war, and, secondarily, of providing some measure of economic justice between nations. Enough of a realist to recognize the impossibility of the League of Nations ever fulfilling such a global role in restricting the sovereignty of individual states, Russell then speculated about which country might become sufficiently powerful to impose such an authority on the rest of the world. His candidate for this role was, not surprisingly, the United States.[28] His proposal for American world domination rested in part on the role of American "high finance." Although he acknowledged that he would "incur the displeasure of most socialists" for saying it, he called high finance "the sanest and most constructive influence in the Western world,"[29] and he said that internationalism at the moment was more important than socialism.[30] An economic drive toward internationalism, Russell felt, might lead to the creation of a few, large self-subsistent states that had only "trivial" commercial contacts with each other, therefore all but eliminating the economic causes of conflict and rendering war something from which no one might profit. Relieved of the burden of defending themselves against aggressors, states would keep only the armed forces needed to maintain internal order among the "subordinate nationalities," which would be allowed autonomy in everything but foreign policy and the control of trade within the state.[31] In time, Russell felt, with the growth of large states and the resulting decrease in the importance of international economic relations, the causes of war might be removed.[32]

For all that he thought American high finance might help prevent the Next War, Russell had little use for the continuation of nineteenth-century capitalism into the Machine Age. Ultimately, the risk of what was called the "class war" was as much of a threat to the old savage as was national rivalry. He considered private property a holdover from an agrarian society, continued and maintained by the power of the state. Private property thus depended upon the ability of the state to protect itself, and the property of individuals depended upon the legal rights granted to them by the state.[33] Because of the collective labor required for large-scale industrial production, the individual worker had become dependent on the people who provided the capital to sustain the industry, and who directed its production. The inequality of their respective positions in this relationship rendered ineffectual the application of liberal ideals to a capitalist industrial society.[34] Because industry tended to benefit from increased size and the coordination of its different aspects (such as what we now call the "vertical integration" of the supply of raw materials, the manufacturing of commodities, and the means of their distribution), Russell pointed out how the power of the "trusts" had already matched or exceeded those of the State, thereby turning the State into what the socialists and syndicalists were wont to call "an organ of the capitalists."[35] This had serious consequences, he believed, which would inevitably result in the "class war" that was a likely candidate for the Next War.

As industrialism made society more interdependent, it increased the power of the state, giving it "a wholly new power" over the lives of its citizens. At the same time, education had contributed to the growth of a political democracy that was rendered "almost worthless" by large-scale economic organization and the concentration of power in the hands of a few capitalists. Calling the private capitalist "an unduly anarchic survival" in the industrial era, "preserving for himself alone a form of liberty which the rest of the community has lost," Russell therefore asserted that capitalism "rouses an opposition which must in the end destroy it."[36] The question for him was whether or not labor would be able to establish socialism on "the ruins of capitalism," or whether industrial civilization as a whole would be destroyed in the course of the struggle.[37] He was not an advocate of socialism by revolution, and therefore he regarded the class war as something to be avoided.[38]

This assessment was strengthened by Russell's experience of Bolshevik Russia. He had a conservative's fear of social instability, especially in a democracy, and he considered the cost of revolution far too high for the ostensible benefits that awaited on the far side of such a conflict. For him, "only peace and a long period of gradual improvement" could bring about a reduction in the inequality of power.[39] Russell rejected the need for a class war to bring about the replacement of capitalism by communism. The emphasis he saw upon hatred and vengeance in communist thought would make that advent so catastrophic there was no assurance that the society which survived it would be amenable to socialist government.[40] Socialism in Great Britain and in the United States, he felt, would thus find its best opportunity in persuading the masses democratically to overturn the capitalist system.[41]

Russell's own version of "socialism" was intermingled with aspects of liberal thought and thus seemed to have more of a utopian, rather than a practical character.[42] For example, he considered democracy to be the only means of reducing government interference with individual liberty in a socialist state.[43] In *Proposed Roads to Freedom,* Russell had evaluated the possibilities presented by three rival theories—socialism, anarchism, and syndicalism—whose proponents heralded the dawn of the post–capitalist era. In the end, he favored an amalgam that he called "guild socialism."[44] Russell's understanding of socialism is best viewed as a philosophical response to the dilemmas posed by industrial capitalism, however, rather than in terms of its relation to some of the other socialist programs of its day. Throughout the interwar period, his "retreat from Pythagoras" had not progressed to the point where he was able to relate his philosophical socialism to the pragmatic realities of Anglo-American political life. His vague depiction of a socialist society in which people would pursue creative rather than possessive impulses depended upon the rather wistful idea that people had to become more concerned with pursuing their own happiness rather than with pursuing the misery of others.[45] It was a vision of a society that he wished people could accept, rather than one that they were likely to institute.

SCIENCE, TECHNIQUE, AND SOCIETY

In assessing the reasons for the struggle between industrial civilization and humanity, Russell had to confront the fact that modern industrialism had made obsolete the nineteenth-century conception of science as the pursuit of truth and knowledge. Science and industry were intertwined in the Machine Age; science was an expression of social, economic, and political power, a tool of capitalists and communists alike. Russell tried to represent science as something more than the instrumental, utilitarian, and mechanistic enterprise that many people in the interwar period understood it to be. In order to do this, he first had to repudiate, as something both philosophically inadequate and socially destructive, the popular representation of science as technique which underlay the conflict between the individual and the machine.

During the interwar period, Russell came to realize the centrality of "power" to the problems and prospects of industrial civilization. He disagreed with the Bolsheviks' idea there was one evil at the root of all others, and he said that if he had to choose the greatest political evil, it would be the "inequality of power" instead of the inequality of wealth.[46] Thus while the applications of science brought power, both military and economic, no revision of the existing political structures that failed to address the unequal distribution of that power would ultimately be successful. In the same vein, the moral progress of the "old savage" also required a development in his ability to wield such power wisely, to the benefit of others beside himself. Yet Russell noted how individual freedom had been lost as a result of the inequalities of power in Western society, inequalities that an industrialism dominated by capitalism and nationalism was bound to increase.

Russell considered science both the foundation of whatever hope there was for the future of industrial civilization and the agent of whatever destruction seemed imminent:

> It is science, ultimately, that makes our age different, for good or evil, from the ages that have gone before. And science, whatever harm it may cause by the way, is capable of bringing mankind ultimately into a far happier condition than any that has been known in the past.[47]

He realized that the prospects of industrial civilization depended on humanity regaining rational control over the new forces that science had unleashed, but he was not optimistic about the possibility. It was Icarus, the son of Daedalus, who rashly flew too high, had his wings melted, and fell to his death, which was for Russell an appropriate emblem for the future of science. He warned in *Icarus, or The Future of Science* (1925) that, "science will be used to promote the power of dominant groups, rather than to make men happy."[48] Science had generated an increase in the "intensity of organization" that was essential for industrial society, but blame for the ambivalent prospects of that society was to be attributed to the individuals who controlled it, he said, rather than to science itself: "Science enables the holders of power to realize their purposes more fully than they could otherwise do. If their purposes are good, this is a gain; if they are evil, it is a loss."[49] This led him to hold out hope for the future of civilization on the one hand, while dashing it on the other hand, saying "only kindliness can save the world," but that no one had a reason to be kind. The popular understanding of science, as technique or as something more than technique, was instrumental in creating the moral framework of whatever new civilization in which the old savage hoped to live.

The Scientific Outlook (1931) was Russell's attempt to articulate what he had earlier referred to as the "scientific temper of mind" necessary to the survival of humanity in the midst of industrial civilization. It was intended to deal solely with the interrelationships of scientific method, scientific technique, and the scientific society that was in the process of being created. He set out three goals: first, to discuss the nature and scope of scientific knowledge; second, to depict the increased power to manipulate nature, which is a product of scientific technique; and third, to suggest what changes in society and in traditional institutions would result from the new kind of organization that scientific technique required.[50] From the beginning, however, Russell made it clear that he was most concerned with scientific technique, pointing out that science-as-knowledge was gradually being pushed into the background by science-as-technique. While science as the pursuit of truth was the equal of art, he said, science as the power of manipulating nature had "a practical importance to which art cannot aspire," even though it may have "little intrinsic value."[51] Further, the influence of science on society was increasing, affecting everything from forms of economic organization to the powers of the State, and to the nature of family life. Though science-as-knowledge underlay the other two forms, Russell (like his audience) was aware of the immediate need to grapple with the implications of science-as-technique if industrial society was to continue. The new power, derived from science and wielded by humanity, was not in itself good or bad, he said, but if a scientific civilization was to be good rather than bad, the increase in knowledge attributable to science had to be accompanied by an increase in the wisdom concerning its use.[52] Yet, Russell warned, science did not provide wisdom, nor did it provide any guarantee of progress in the absence of wisdom. While *The Scientific Outlook* was ostensibly concerned only "with science rather than with wisdom,"[53] throughout the book, as in the rest of Russell's writing from the interwar period, the question about what the "old savage" should do with the power scientific technique had given him was a recurrent theme.

In *The Scientific Outlook,* Russell maintained that knowledge itself was not the source of danger, for knowledge was good and ignorance was evil; nor was power "in and of itself" the source of danger for society. Instead, he asserted, what was dangerous was "power wielded for the sake of power, not power wielded for the sake of genuine good."[54] The leaders of the modern world, because they were "drunk with power," saw in their ability to do what "no one previously thought it possible to do" sufficient reason for doing anything they desired.[55] Russell, however, asserted that power was not an end in itself, but only "a means to other ends," and until those ends were remembered, science would not be able to do what it could "to minister to the good life."[56]

Two sections of *The Scientific Outlook,* "Scientific Technique" and "The Scientific Society," dealt with specific aspects of the application of science to society, both present and future. Russell first dealt with the development of scientific technique, distinguishing it from the techniques of arts and crafts by "its most essential characteristic," the fact that it proceeded from experiment and not tradition.[57] While this spirit of experiment was difficult to maintain in successive generations—for whom techniques obviously become a product of tradition rather than experiment—it was responsible for the dramatic increase in the last 150 years in "the power of man in relation to his environment" over any previous civilization. Russell then briefly outlined some of these applications of scientific technique in the areas of machinery, biology, physiology, psychology, and finally in society itself. His portrayal of the application of technique to social questions, however, was intended to set up the section in which he depicted "the scientific society."

Russell conceived of the scientific society not only as one that employed the best techniques in production, education, and propaganda but also as one that had been created "deliberately with a certain structure in order to fulfill certain purposes."[58] Unlike past societies, which grew by "natural causes" and without much conscious planning, the scientific society would be one in which nothing possible was left to chance. This society, moreover, would be worldwide, taking full advantage of the resulting benefits of economic and political organization—like the elimination of warfare—which were so great that Russell called them "an essential condition for the survival of societies possessing scientific technique." He thought the argument for survival was so compelling as to render "almost unimportant" whether or not individual life in such a unitary World State would be more satisfactory than at present.[59]

In this utopia, there would be no poverty, no unemployment, and "almost all that is tragic in human life" would be eliminated, even to the point that death before old age would be unusual.[60] Russell said he did not know whether humanity would be happy in this "Paradise," but he suggested biochemistry might reveal "how to make any man happy, provided he has the necessaries of life."[61] Through sports or some other activities, people might find a safety valve for the "anarchic" side of human life, which had become completely regulated by the state. There would be a universal language ("either Esperanto or pidgin-English") into which all literature would be translated, as long as it did not promote values or ideas contrary to the public peace. All of these controls, Russell said, were necessary because of the power attributable to science, which has increased the need "for restraining destructive impulses." If this scientific society was to survive, human beings had to become "tamer than they have been," and the notion of the "splendid criminal" had to be replaced by a more submissive social ideal.[62]

In the end, it was the question of the new relationship between the individual and society science and industrialization seemed to require that seemed to pose the greatest difficulty for Russell. Individual liberty in a variety of forms had to be curtailed or eliminated altogether in favor of the collective good of the scientific society. So the investment of capital according to individual whims was not "socially defensible."[63] Private housing was equally absurd—

children would be cared for in a communal or collective environment, freeing up for useful employment the women who had neither the inclination nor the temperament to take care of them. The freedom exercised by young people to choose their own profession would be taken away, and they would be assigned jobs according to their abilities, which might be established as early as three years of age. The population would be controlled, both as to its quantity and its "quality" in determining who should be allowed to have children, and how many of them. All of these things, and other examples of the regulation of society, would be a product of "scientific government," which itself would grow increasingly oligarchical, as the population would be sorted into a governing class and the equivalent of the "working class."

By "scientific government," Russell meant government according to scientific principles, not government by scientists. The scientifically organized society would do away with an economic structure derived from feudalism, and the "hostile idealisms," such as Christianity, that affected social ethics by emphasizing the individual over the collective good. The "new ethic" gradually growing out of scientific technique, Russell said, would have "little use for guilt and punishment," preferring instead to make individuals suffer for the good of the community "without inventing reasons purporting to show that they deserve to suffer."[64] This society would be "ruthless" according to present morality, he said, but the change would come about "naturally through the habit of viewing society as a whole rather than as a collection of individuals."[65] Observing that this mentality was already acceptable in time of war, Russell felt that the "scientific idealists of the future" would have no problem applying it to social organization in peacetime as well.

How the world would end up being governed in this fashion was not a pleasant prospect. Russell was at his gloomiest in speculating on the impending destruction of Europe, and its reconstruction by the United States, or some other country sufficiently isolated from the inevitable demise of European capitalism. The society that emerged from the rubble would be run by an oligarchy of experts and would hopefully be more stable than what had preceded it. Worldwide organization should mean the global control of agriculture and the supply of raw materials, as well as the elimination of nationalism. This, along with a monopoly on military might, should help reduce the likelihood of further global warfare. Scientific innovation would probably cease, he thought, because the hierarchy of experts would be dominated by tradition rather than experiment. New ideas would be seen as potentially threatening to the status quo.

Some of the most disagreeable aspects of this scientific society were those Russell saw in the future of the family, how social control over reproduction and the education of children would fundamentally alter previous patterns of life. "Scientific breeding" meant eugenic control of the population and the application of scientific technique to the production of the right kinds of individuals, perhaps suited to certain occupations by selecting their heredity. Education would be rigidly controlled, with each child tested throughout the process (like a manufactured item) to ascertain where in the society he or she would fit. The result of this kind of state control of children, moreover, would mean radical changes in family life. Russell thought that the sentiment of paternity, for example, would "disappear completely," and if children could be separated from their mothers, perhaps after inducing a premature birth, maternal feeling might have "little chance to develop."[66] In this scientific society, there would be pleasure, he said, but there would be no joy. He likened it to a kind of ascetic existence, alleviated by the pleasures that might come from "injections and drugs and chemicals," or "new forms of drunkenness," which might induce the population "to bear whatever its scientific masters may decide to be for its good."[67]

Obviously, this kind of society was anything but desirable to Russell. His portrayal, however, was within the realm of practical politics, and whether he liked the idea or not was immaterial:

> The future which I foresee is, to begin with, only very partially in agreement with my own wishes. I find pleasure in splendid individuals rather than in powerful organizations, and I fear that the place for splendid individuals will be much more restricted in the future than in the past.[68]

Only at the end of this depiction of his scientific Nirvana did Russell betray his own feelings about what Aldous Huxley would go on to call the "brave new world":

All these are possibilities in a world governed by knowledge without love, and power without delight. The man drunk with power is destitute of wisdom, and so long as he rules the world, the world will be a place devoid of beauty and joy.[69]

Thus he began his final chapter of *The Scientific Outlook*, "the scientific society which has been sketched . . . is, of course, not to be taken altogether as serious prophecy."[70] He intended it as a depiction of the world if "scientific technique [should] rule unchecked." It combined ideas that would elicit both positive and negative responses from his reader, emphasizing certain elements of human nature to the detriment of others. He considered "the impulse towards scientific construction" to be "admirable," as long as it was not pursued to the exclusion of other values in human life, in which case, it could easily turn into the kind of "tyranny" he had described.[71] That tyranny involved the mechanical conception of society in which utility and organization were dominant virtues.

THE UTILITARIAN OUTLOOK

The diminution of individual liberty concerned Russell, for as much as he could understand its necessity in an industrial society, he also felt "special measures" had to be undertaken to prevent it from being eradicated entirely by the utilitarian attitude that characterized industrialism. This had been one of the main themes in his earlier pamphlet, *Political Ideals* (1917).[72] In it, Russell held that political ideals needed to be expressed in terms of the lives of individuals rather than in terms of the collective group to produce a truly "good" society.[73] Good and bad societies may be evaluated only on the basis of how the individual is treated. What was more, he believed that minority opinions and ideas within a society were actually the source of constructive change and not just of instability, for a tyranny of the majority, whether intellectual or political, led to stagnation and not progress.[74]

Russell thought the utilitarian frame of mind that industrialism tended to produce was both unfortunate and unnecessary. In *Prospects,* he depicted this attitude as "the tendency to value things for their uses rather than for their intrinsic value."[75] Because most individuals build machinery which, in its turn, produces commodities to consume, Russell said this made people more utilitarian than artistic, as nothing they made had in itself "any direct human value." As he put it, the journey from means to ends was so long and involved that most people could not see the ends at all and came to think production was the only thing of importance: "Quantity is valued more than quality, and mechanism more than its uses."[76] He then warned against anyone regarding humanity as a means of producing commodities rather than seeing commodities as "a subordinate necessity for liberating the non-material side of human life."[77]

The utilitarian outlook was for Russell one of the forces causing the decline of romance and art, but he also said that its effects went deeper, upsetting humanity's dreams of a better world, and "their whole conception of the springs of action."[78] Against proponents of a mechanistic conception of society, he declaimed:

> Man's true life does not consist in the business of filling his belly and clothing his body, but in art and thought and love, in the creation and contemplation of beauty and in the scientific understanding of the world. If the world is to be regenerated, it is in these things, not only in material goods, that all must be enabled to participate.[79]

Science as knowledge was imperiled by its own success. Russell felt that the scientific outlook needed to include more than the mere manipulation he associated with scientific technique. Recent history illustrated a shift in the understanding of science from contemplation to manipulation, and he wanted to remind his audience of the importance of "knowledge for its own sake" in the face of the overwhelming influence of "knowledge as power."

The love of knowledge, which he identified as essentially being responsible for the development of science, was itself the product of "a twofold impulse," in which knowledge of an object is sought because "we love the object or because we wish to have power over it."[80] The contemplative impulse unfortunately had been superseded by the practical impulse, he asserted, replacing love with power, and this fact was reflected both in industrialism and in government. In philosophy he observed that the same attitude was reflected in pragmatism and instrumentalism, which he characterized as holding "that our beliefs about any object are true insofar as they enable us to manipulate it with advantage to ourselves."[81] It was a "governmental view of truth" that these philosophies, and scientific technique in general, offered to society, seemingly without limit.[82]

SCIENCE AND VALUES

Science involved more than its applications for Russell, however, and no theory of truth that was instrumental was worthy of the name. Science-as-knowledge yielded the applications upon which industrial civilization was being built, but those applications could never be a substitute for what science revealed about the nature of the universe. He granted that modern science had "baffled" the "lover of nature" while rewarding the "tyrant over nature," however, and he said that this imbalance had to be redressed if society was to be anything more than a sadistic tyranny, in which power-knowledge has been substituted for love-knowledge:

> The scientific society of the future as we have been imagining it is one in which the power impulse has completely overwhelmed the impulse of love, and this is the psychological source of the cruelties which it is in danger of exhibiting.[83]

Considered as technique, science gives humanity an unlimited power over its environment that is "quite independent of its metaphysical validity," but it is a power that can only be wielded "by ceasing to ask ourselves metaphysical questions as to the nature of reality."[84] These are the questions that a lover would ask, thus it is only insofar "as we renounce the world as its lover that we can conquer it as its technicians." Yet, he said, "this division in the soul is fatal to what is best in man," and it was "the fundamental reason why the prospect of a scientific society must be viewed with apprehension."[85] The scientific society "in its pure form" was "incompatible with the pursuit of truth, with love, with art, with spontaneous delight, with every ideal that men have hitherto cherished."[86]

Russell maintained that the pursuit of power was itself hollow, because it never comes to an end; the person who pursues power is unable to contemplate what she or he has acquired, only to search for some further opportunities for "fresh manipulation."[87] He contrasted the futility of this attitude with "the lover, the poet, and the mystic," saying that they find "a fuller satisfaction than the seeker of power can ever know, since they rest in the object of their love."[88] Thus "the satisfactions of the lover" should be given a higher place among the ends of life than "the satisfactions of the tyrant." For science, it meant that if science could bring people the opportunity to experience these other things, the power it wields will have been "wisely used." If, however, science is used to take out of life "the moments to which life owes its value," then "science will not deserve admiration, however cleverly and however elaborately it may lead men along the road to despair."[89] Russell concluded:

> The sphere of values lies outside science, except insofar as science consists in the pursuit of knowledge. Science as the pursuit of power must not obtrude upon the sphere of values, and scientific technique, if it is to enrich human life, must not outweigh the ends which it should serve.[90]

Scientific government, then, had "to make life tolerable for those who are governed," not merely to afford pleasure to those who govern. Scientific technique could no longer be allowed to form "the whole culture of the holders of power."[91] What the manipulator of science needed to realize, Russell said, was that knowledge and feeling were equally essential as "will" to the good life of the individual and the community. He upheld the value of knowledge for its own sake, but he also asserted that "the life of the emotions" was even more important than knowledge, for "a world without delight and without affection is a world destitute of value."[92] The wielders of power had to remember the truths of other generations, he said, for "not all wisdom is new, nor is all folly out of date":[93]

> The new powers that science has given to man can only be wielded safely by those who, whether through the study of history or through their own experience of life, have acquired some reverence for human feelings and some tenderness towards the emotions that give colour to the daily existence of men and women.[94]

Russell concluded *The Scientific Outlook* with the observation that science had overturned humanity's subjection to nature but had proceeded in turn to enslave humanity to its baser instincts. Part of the "scientific outlook," therefore, had to be "a new moral outlook . . . in which submission to the powers of nature is replaced by respect for what is best in man."[95] Scientific technique was dangerous when this respect was lacking, but the situation was not hopeless; "hope for the future," he said, was "at least as rational as fear."[96]

FROM OUTLOOK TO WORLDVIEW

If the problem of the "old savage and the new civilization" was at heart a moral one, in that moral capacity had not kept pace with technological development, then there needed to be some basis on which to establish a scientific morality appropriate to the new demands that science and technology had placed on the individual in modern society. In the two volumes of *The Philosophy of Civilization,* which were published in 1923 (the same year in which Russell's *Prospects* appeared), Albert Schweitzer tried to attribute the current state of morality to the absence of an appropriate "worldview."[97] He saw the Great War and its moral aftermath as the result of European society having given up the "worldview" of the Enlightenment. He attempted to provide the framework for a new *Weltanschaaung* appropriate to the Machine Age, founded on what he called the "reverence for life."

While Russell was critical of Schweitzer's second volume, *Civilization and Ethics,* his own attempts to articulate "the scientific temper" or "the scientific outlook" involved an attitude similar to Schweitzer's. Although he said the book was of "considerable importance" and deserved "to be read with care," he disparaged the idea "that our views on ethics must be dependent upon our views as to the nature of the world."[98] Yet Russell himself subsequently struggled, in works of his own, with the relationship between different views of the nature of the cosmos and the ethical character of the society created by modern science. He was biting in his dismissal of Schweitzer's idea that a change in social ethics might lead to a change in practice,[99] but, in *The Scientific Outlook* (1931), Russell later advocated a similar alteration in "moral outlook" to change the way scientific technique was exercised in modern society.

In fact, I suggest that Schweitzer's argument had more of an influence on what Russell wrote in the interwar period than would first appear to be the case. There is a similarity between *The Philosophy of Civilization* (1923) and *Principles of Social Reconstruction* (1916) which, even if Schweitzer had not read Russell's book, illustrates that they had earlier shared a certain perception of what needed to be done. Part of the sharpness of Russell's rebuttal of Schweitzer's second volume could have been due to his realization, after the War, that a more pragmatic response was needed to the problem of the old savage than he had earlier presented in *Principles*.

More significantly, I think that Schweitzer's influence relates to Russell's struggle, during the period 1923–1931, to articulate his "scientific outlook" in terms other than those established by science-as-technique and the utilitarian attitude reflected in the mechanistic conception of society. I suggest Russell took his own advice when he said in his review that Schweitzer's book was of "considerable importance" and deserved "to be read with care." Schweitzer's understanding of "worldview" illuminates both Russell's "scientific outlook" and similar expressions found in the work of their contemporaries during the interwar period.

Schweitzer defines "worldview" as "the content of the thoughts of society and the individuals which compose it about the nature and object of the world in which they live, and the position and destiny of mankind and of individual men within it."[100] In terms of renewing civilization and halting its decay, Schweitzer says that the individual is the "sole agent" for such a renewal, and that no social ethic alone will accomplish this reversal:

> Social ethics without individual ethics are like a limb with a tourniquet around it, into which life no longer flows. They become so impoverished that they really cease to be ethics at all.[101]

Like Russell, he rejects the kind of society that is organized for the benefit of the group, without valuing the individual:

> The essence of humanity consists in individuals never allowing themselves to think impersonally in terms of expediency, as does society, or to sacrifice individual existences in order to attain their object. The outlook which seeks the welfare of organized society cannot do otherwise than compromise with the sacrifice of individuals or groups of individuals.[102]

Russell must have found all of this very congenial, and it certainly squares with what he would later write in *The Scientific Outlook*. Yet the problem for Russell stemmed from the fact that Schweitzer took these ideas to demonstrate that individual ethics is the product of a personal worldview, and that social ethics must be the result of a worldview shared by the individuals in a society. While Russell wanted something with which to oppose the mechanistic worldview on behalf of the individual, he was not comfortable with the obvious association between Schweitzer's "worldview" and a mystical or religious outlook. For members of a society to share the same worldview, there would need to be some universal basis for asserting its validity, one that would need not have its roots in a scientific conception of truth. If a social ethic was dependent on such a universal worldview, and if it were to be "true," then that would entail another means to the acquisition of truth about the universe than through science.

At the end of *The Scientific Outlook,* Russell flirted with a metaphysical outlook in science as an alternative to the mechanistic outlook that he deplored. Dissatisfied with the representation of science-as-technique alone, which led to a conflict between the individual and the machine in society that the individual was destined to lose, Russell wanted to promote a view of science consonant with an outlook that was something other than mechanistic and instrumental. Whether that outlook was "moral," humanistic, or metaphysical, it entailed an understanding of truth that was not governmental and a promotion of values not determined solely by their social utility.

The popular identification of science with technique and the extent and rapidity of social change attributable to technology and not to philosophy pushed Russell to consider what else science might mean to people living in the Machine Age. The relationship between knowledge and values, or between knowledge and the wise use of that knowledge, was indisputably tied to the problem of the "old savage in the new civilization." There could be no workable embargo on knowledge, nor could there be a moratorium on scientific research; what was needed was a means of encouraging correlative moral progress on the part of the individuals who would be presented with this knowledge. What was meant by moral progress, and how it might be accomplished, was less obvious than the need for it; if the old savage was to survive in the new civilization, the only alternative to individual moral development was the coercion of individuals by the state, which would inevitably be according to the interests of those who wielded power most effectively.

In rejecting a social ethic derived from the operations of science, Russell thus found himself having to discover an alternative that reflected something more than the utilitarian outlook he found in the mechanistic conception of society. He expressed that "something more" in terms of a moral outlook, aesthetic values, and a sense of mystical experience in the pursuit of scientific knowledge, recognizing the importance of "science-as-metaphysic." At the end of *The Scientific Outlook,* Russell therefore moved away from the rational approach to scientific knowledge that would conventionally characterize the application of the scientific method to society. The ethical dilemmas posed by scientific technique for the future of "the old savage in the new civilization" had caused him to formulate, however tentatively, a "scientific outlook" in response, which incorporated moral and aesthetic elements and in particular promoted the importance of the individual over against the machine.

For the old savage to survive in the new civilization made possible by science and technology, however, this new outlook had to lead to a social ethic appropriate to the Machine Age. While he deplored the effects of the Machine Age on the individual, Russell also saw the need for the "old savage" to be restrained. The choice for him was between a new social ethic based on the external constraints available to society in a scientific civilization and the moral consensus of individuals who would choose collectively to change their destructive behavior. As much as Russell disliked the idea of a utilitarian ethic, based on power derived from the applications of science, he found himself returning to this idea when he realized the various implications of an outlook based on something else. Russell's antipathy to the mechanistic outlook coincided with the ideas of those who saw, in the philosophical implications of the new physics, a place for religion in the modern world. Although *The Scientific Outlook* concluded with Russell's call for a moral outlook, earlier chapters of the book had lampooned the metaphysical speculations of Alexander Eddington and James Jeans, prominent physicists who had both used recent discoveries in physics to suggest the existence of some other form of knowledge than the scientific, and a corresponding sense of purpose in the universe. Russell was ultimately unable to accept the resulting role that religion or metaphysics would then be given, either in ethics or in the worldview of a scientific society in which such a perspective was valued. His rejection of religion in a world governed by science eventually led to his inability to find other grounds than the utilitarian exercise of power on which to establish a "moral outlook" in the new civilization.

Chapter 3

Religion, Metaphysics, and Meaning

Religious and metaphysical questions were matters of frequent public debate in the popular press during the interwar period. Russell and his contemporaries were confronted by existential questions that arose from the catastrophe of the Great War and from the new discoveries in the physical and social sciences. Will Durant's ambitious volume *On The Meaning of Life,* for example, began with a rather brash letter that he sent to a wide variety of influential people in 1931. He asked his correspondents ("famous contemporaries here and abroad for whose intelligence I have high regard") to interrupt their work and "play the game of philosophy with me."[1] The question he posed to them was one "which our generation, perhaps more than any, seems ready to ask and never able to answer—What is the meaning or worth of human life?"[2] Perturbed by the current disillusionment of the post-War generation and the apprehension about a future controlled by technology ("which has almost broken the spirit of our race"),[3] Durant concluded: "Life has become, in that total perspective which is philosophy, a fitful pullulation of human insects on the earth, a planetary eczema that soon may be cured."[4] He blamed the pursuit of truth for such an appalling prospect, saying "it has not made us free, except from delusions that comforted us and restraints that preserved us."[5] Truth, he concluded, was not beautiful, and has "taken from us every reason for existence except the moment's pleasure and tomorrow's trivial hope."[6] This, he said, "was the pass to which science and philosophy have brought us."

Durant's correspondents were asked to tell him "what meaning life has for you, what keeps you going, what help—if any—

religion gives you, . . . where you find your consolations and your happiness, where in the last resort, your treasure lies."[7] His book contained answers from diverse personalities such as Will Rogers, Havelock Ellis, Theodore Dreiser, and Jawaharlal Nehru.[8] Durant quoted Russell's reply under the heading of "Skeptics and Lazybones," introducing him as "the Bad Boy of England, scandalizer of continents, and prospective terror of the House of Lords."[9] Russell was to the point:

> Dear Mr. Durant,
>
> I am sorry to say that at the moment I am so busy as to be convinced that life has no meaning whatever, and that being so, I do not see how I can answer your questions intelligently. I do not see we can judge what would be the result of the discovery of truth, since none has hitherto been discovered.[10]

It is no accident that Russell associates "the meaning of life" with the general inability to discover truth. Questions about meaning involve questions about epistemology and truth: what we know, what can be known, and whether we have any reason to think it true are all aspects of any attempt to assign either meaning, or meaninglessness, to human existence. Thus we cannot reach a conclusion about what Russell thought human existence might mean without examining his views on the philosophical implications of the universe uncovered by the new physics. Nor can we understand what Russell meant by a "moral outlook" (the need for "values" other than the instrumentalism inherent in scientific technique) without first clarifying what he thought about religion and its relation to science and morality in the interwar period.

In any discussion of Russell and religion, difficulties abound. Merely to mention Russell and religion in the same sentence is to risk philosophical confusion or provoke the ire of those who accept his dismissals of religion at face value. To avoid both perils, it is wise to clarify the approach I have taken to understanding his perception of religion, science, and morality in the interwar period.

First, Russell's repeated rejection of Christianity must be respected.[11] Whatever his religious beliefs, he was not, at any point, converted from skepticism to belief like a Malcolm Muggeridge. At the same time, however, while Russell and his contemporaries knew of "the varieties of religious experience,"[12] religion for them was still conceived in terms of Western Christianity. Even though Russell had the experience of visiting China, apart from the odd reference in his writings to Buddha or Lao Tse, his critique of religion was overwhelmingly expressed in terms of Western Christianity. Even if Russell vehemently rejected all aspects of Christianity, it still set the terms of the debate for him.[13]

Second, there is no good reason to assume a lifelong consistency and continuity in Russell's view of religion that is not apparent in his ideas on other subjects. There is thus the need for a chronological representation of what he wrote on religion. When we apply the idea of the Great War as a watershed in Russell's personal life to his views on religion and metaphysics, there is a decided difference between pre–War pieces such as "The Free Man's Worship" and "The Essence of Religion" and what he wrote during the interwar period. While the change in Russell's tone between "The Essence of Religion" (1912) and "Why I Am Not a Christian" (1927) likely had something to do with the end of his personal relationship with Ottoline Morrell, it is just as likely to have been the result of the same disillusionment with the traditional forms of Christianity in Western society that so many of his contemporaries experienced as a result of the Great War.

This leads to the third point, which is my assessment of what Russell thought about religion during the interwar period. I contend that he was more concerned with the character of modern society than with the specific beliefs of its members. Christianity, whatever its personal appeal or metaphysical absurdities, had, as a social institution, demonstrated its inability either to prevent the Great War or to direct the morality of post–War society in a direction that would enable it to avoid the Next War. As a result, Christianity joined the list of archaic institutions and ideologies incapable of dealing with life in the Machine Age. Thus, for Russell, the important question after the Great War was not the validity of a Christian metaphysic but whether such a metaphysic could yield anything of value for modern society. The dilemma posed by "the old savage in the new civilization," and the need for a "moral outlook" in an age of scientific technique, meant for Russell that Christianity needed to be assessed in terms of its social utility, not its expression of abstract truth. As Russell retreated further and further from Pythagoras, personal belief (his own or someone else's) became less significant than the actions that resulted from the collective expression of those beliefs. The prospect of a Next War, in which religion (or institutional Christianity) was equally complicit, was considered by Russell to be grounds enough for his trenchant dismissal of its validity.

Thus we need to distinguish between the personal and the social views of religion in Russell's work. Whatever his other inclinations about the truth or validity of religion, during the interwar period the problem of meaning for Russell could not be separated from the dilemmas posed by trying to ensure the survival of "the old savage in the new civilization." Any purely personal morality or religion was not an adequate response to the problems of modern society if humanity was to survive in the Machine Age. His most serious treatment of the subject was therefore in social and institutional terms, making questions about the existence of God, for example, incidental to the problems of formulating a workable social ethic.

While questions about the truth of religious doctrines were more significant to him at an earlier stage in his "retreat from Pythagoras," the nature of knowledge, and how it could be acquired, also was related to Russell's negative perception of religion in the interwar period. In *Religion and Science,* he wrote that science "encourages abandonment of the search for absolute truth," replacing it with a concern for "technical truth," which belonged to any theory capable of being used in "inventions or in predicting the future."[14] Technical truth was perfectible, and thus "a matter of degree"; its effect on epistemology also made knowledge more tentative: "'Knowledge' ceases to be a mental mirror of the universe, and becomes merely a practical tool in the manipulation of matter."[15] Thus whatever religion claimed to reveal or thought it represented, as far as Russell was concerned, it could neither be considered knowledge nor truth.

RUSSELL'S PHILOSOPHY OF RELIGION

Russell did not have a "philosophy of religion" specific to the interwar period, in the sense that he had worked out a rational, systematic account of the nature and function of religion in abstract, personal, or social terms. Instead, there was a patchwork of biting asides, vituperative comments, perfunctory dismissals, and rhetorical posturing. It is no wonder that, in his notable discussion of Russell's philosophy of religion, Edgar Brightman dealt almost exclusively with what Russell wrote before the end of the Great War.[16]

Russell's essay "Has Religion Made Useful Contributions to Civilization?" reflects the short shrift that religion in general received from him during the interwar period. Agreeing with Lucretius, he called it "a disease born of fear and . . . a source of untold misery to the human race."[17] Its only two positive contributions, he trumpeted, were fixing the calendar and causing Egyptian priests to try to predict eclipses. He criticized the vague use of the term *religion* by his contemporaries who, "under the influence of extreme Protestantism, employ the word to denote any serious personal convictions as to morals or the nature of the universe."[18] Saying this was "quite unhistorical," Russell claimed religion was "primarily a social phenomenon."[19] To understand the social influence of Christianity, he said, one had to look at the Church, not at the teachings of Christ. He identified the two main objections to religion as being intellectual and moral:

> The intellectual objection is that there is no reason to suppose any religion true; the moral objection is that religious precepts date from a time when men were more cruel than they are, and therefore tend to perpetuate inhumanities which the moral conscience of the age would otherwise outgrow.[20]

He refused to separate the truth claims of religion from its social function and proceeded to dismantle various Christian beliefs, asserting the three human impulses which characterize religion are "fear, conceit, and hatred."[21] Russell then tried to make a case for the eradication of these impulses through psychology and "our present industrial technique,"[22] applied through education. Calling the teaching of religion the chief obstacle to universal happiness, he concluded:

> Religion prevents our children from having a rational education; religion prevents us from removing the fundamental causes of war; religion prevents us from teaching the ethic of scientific cooperation in

place of the old fierce doctrines of sin and punishment. It is possible that mankind is on the threshold of a golden age; but, if so, it will first be necessary to slay the dragon that guards the door, and this dragon is religion.[23]

The unsophisticated nature of Russell's critique and the extravagant blame heaped upon "the dragon of religion" for all of the evils of civilization illustrate the problems inherent in discussing his philosophy of religion. With a good rhetorical wind blowing, Russell often sailed further in a particular direction than he might otherwise have done. Having criticized some of his contemporaries for their vague use of the term *religion,* passages such as this one demonstrate that Russell himself was not immune from the same criticism.

More significantly, Russell distinguishes between the personal and social views of religion, favoring its perception in terms of society over its meaning in personal experience. This is the single most important difference between what he wrote on the subject before the War and what he wrote on it during the interwar period, and this is where the influence of the Great War may be seen on Russell's view of religion. Apart from the metaphysical absurdities he found in the attempts of his contemporaries to reconcile science and religion, he was much less interested in the individual's search for God during the interwar period than in the ethical implications of beliefs, religious or otherwise. It might be said that the ideas about religion expressed in "The Free Man's Worship" or "The Essence of Religion" had become trivial by the end of the Great War, at least in comparison to the larger questions posed for society by the prospect of the Next War.

RELIGION AS PERSONAL EXPERIENCE

In "The Free Man's Worship" (1903), Russell wrote of the philosophical implications of the cosmos depicted by science up until that point. It was a cosmos without purpose or design, in which the desires and beliefs of humanity were inconsequential, and whose eventual dissolution was certain. As a result, "only on the firm foundation of unyielding despair," he said, "can the soul's habitation henceforth be safely built," in such "an alien and inhuman world."[24] The only gods that exist are those of human creation, and the choice presented to humanity is to worship Force or Power (which "is largely bad"[25]) or to worship Goodness:

> In this lies Man's true freedom: in determination to worship only the God created by our own love of the good, to respect only the heaven which inspires the insight of our best moments. In action, in desire, we must submit perpetually to the tyranny of outside forces; but in thought, in aspiration, we are free.[26]

Therefore, he concluded:

> Let us learn, then, that energy of faith which enables us to live constantly in the vision of the good; and let us descend, in action, into the world of fact, with that vision always before us.[27]

It is evident that the "vision of the good" of which Russell spoke was something akin to a mystical insight but expressed in terms of a Promethean defiance of the meaninglessness of the accidental cosmos in which humanity must exist. There was, throughout the essay, an insistence on the need for transcendence, the need "to burn with passion for eternal things," which was the essence of "the free man's worship." While Russell's prose was overblown, whatever else emerged from this essay, there was a strong sense of the importance of the individual's "religious" or "philosophical" experience of the ultimate futility of the material universe and the insight into the human condition that enabled the "free man" to transcend his despair.

In "The Essence of Religion" (1912), Russell tried to articulate a personal religion freed from the effects of dogma, one that developed the theme of human liberation he had expressed only as a Promethean defiance in his earlier work. Lady Ottoline Morrell approved of his efforts, and certainly what Russell attempted to write reflected his own desire to please her.[28] In "Essence," Russell tried to find, in a hypothetical "religion" that expressed the life of the spirit, a way for humanity to transcend the finite experiences that were inevitably part of brute existence. The awareness of "infinity" that he equates with "the experience of sudden wisdom" is "the source of what is essential in religion."[29] He does not accept a mystical view of this wisdom, however, saying that it is not a product of having revealed a world behind our own but of seeing our own world in a new light. In a letter written on January 3, 1912, Russell said he had had a vision "at rare times of stress or exaltation" that seemed to show him

> that we can live in a deeper region than the region of little everyday cares and desires—where beauty is a revelation of something beyond, where it becomes possible to love all men, where Self as a separate fighting unit fades away, and where all common tasks are easy because they are seen as parts of what is greatest.[30]

This vision, however, alternated with one less uplifting, in which he found that "sorrow is the ultimate truth of life, everything else is oblivion or delusion."[31] Russell concludes that the essence of religion is to be found "in passive submission to the universe" rather than in the Promethean defiance he had advocated in "The Free Man's Worship."[32]

In "The Essence of Religion," Russell discussed, in various ways, the meaning of an individual's religious experience, how the universe and life itself may be perceived as meaningful from a personal standpoint. Purging Christianity of its dogmas, he found three things left that would enable humanity to experience the "mystical union" with the universe, which meant the soul's liberation. All three ("worship, acquiescence and love") were depicted in relation to an individual's religious life, as aspects of religion incomprehensible apart from personal experience, and only debased by the collective religious experience out of which dogma inevitably emerged. The "union with the universe," the subjugation of man's "animal nature" by the soul, was, for Russell, the end and purpose of religion.

These kinds of observations indicate a concern on Russell's part with religious experience, even if as a sense of the ineffable rather than an expression of Christian belief. He did not dispute the existence of revelatory experience, and he admitted to having felt something of the sort himself. Brightman elucidated what he thought Russell believed in this regard, according to his pre-War publications, and Russell generally approved of what Brightman said.[33] There was, however, a significant change in subject matter and in tone between "Essence" and what Russell wrote on religion after the Great War. There is almost nothing regarding religion as a social phenomenon in "Essence" or "The Free Man's Worship."

After the War, not only is Russell's scorn for religion as a social institution pervasive, but little is said about the nature of personal religious experience. The plight of the "old savage" had to be addressed in terms of the "new civilization."

RELIGION AS A SOCIAL INSTITUTION

In the interwar period, Russell was preoccupied with religion, or Christianity, as "primarily a social phenomenon."[34] He was appalled at the way in which the Church abandoned pacifism in 1914 and had supported the Great War as fervently as any other social institution. Scattered throughout his work are references to clergy who urged their congregants to enlist or who criticized Russell for his opposition to the war and to conscription.[35] The realization that the Church in Germany was equally convinced of God's support for its side in the conflict only heightened Russell's cynicism over the application of the dogmas of Christianity to the actions of its adherents.[36]

For Russell, the Church as a social institution was characterized by typical institutional problems relating to the abuse of individual freedom by collective power. He considered it as much of an archaism in the modern world as the other expressions of outdated ideologies that he dismissed as being irrelevant to the needs of the Machine Age. The Church's support for the Great War moreover vitiated its claims to any kind of moral leadership in society and provided a contemporary example of the hypocrisy that Russell saw reflected in its entire history. At least by the end of the Great War, he believed that the Church had no redeeming features, apart from the lives of some individuals within it. This conclusion was a dominant theme in Russell's most celebrated anti-Christian polemic, "Why I Am Not a Christian."

Often cited as a substantial statement of Russell's agnosticism, "Why I Am Not a Christian" is, in fact, a remarkably feeble dismissal of Christianity. In this essay, essentially a public lecture that he gave in 1927,[37] Russell devoted the most space to supposedly refuting various arguments relating to the key dogmas of Christianity. The reputation that this lecture acquired likely had more to do with the fact someone dared to say such things publicly than to the intellectual rigor with which any of these dogmas were addressed, which leads again to the element of rhetorical performance in Russell's work. Both this essay and "Has Religion Made Useful Contributions to Civilization?" originally appeared as pamphlets, published by E. Haldemann-Julius.[38] Having taken such a public stance against "organized religion," Russell achieved a certain notoriety for his opinions, which no doubt helped sell other articles on the subject but which also made it difficult for him to adopt a more moderate stance in his formulation of the scientific outlook.

Russell made it clear that he did not find acceptable the vague definitions of religion common to the interwar period. In 1929, he wrote: "I find a tendency among moderns who think themselves broad-minded to use the word for any serious outlook on life, but this appears to me a historical and literary mistake. Religion is primarily a social phenomenon consisting of rites and beliefs held in common by the community."[39] His definition of religion as a social phenomenon was extremely elastic, however. He considered Bolshevism a religion, by which he meant "a set of beliefs held as dogmas, dominating the conduct of life, going beyond or contrary to evidence, and inculcated by methods which are emotional or authoritarian, not intellectual."[40] Thus, "those who accept Bolshevism become impervious to scientific evidence, and commit intellectual suicide."[41] While religion was responsible for "almost all the major ills" of western society, anyone possessing the scientific temper was just as "fundamentally opposed" to Bolshevism as to the Church of Rome, he maintained.[42] Dogmatic certainty and restrictions on freedom of expression were not compatible with "that temper of constructive and fruitful scepticism which constitutes the scientific outlook."[43]

Russell had no use for creeds or dogmatic expressions of truth that had to be accepted, without adequate proof, by individuals who wished to be part of the Church. He saw the existence of dogma as inimical to the scientific spirit, and any institution that promulgated dogma was essentially "religious" rather than "scientific" in character. He believed in the need for free inquiry and the absence of coercion by any group over what the individual came to regard as true. Thus, he said, the problem was not that the creed of the Church was the wrong one, but that a creed existed at all: "As soon as income, position, and power are dependent upon acceptance of no matter what creed, intellectual honesty is imperiled."[44] Truth was dependent upon investigation, not dogmatic or deductive assertions, and still less on the propaganda fostered by groups within the social structure.

As will be seen in Chapter 5, Russell had other concerns about the relationship between creeds and the Christian church, concerns that related to his conflictual depiction of the relationship between science and religion. He saw the existence of creeds as the main source of that conflict. While science involved an epistemology based on the provisional character of the truth it discovered about the universe, creeds were based on the dogmatic assertion of absolute truth, grounded in revelation rather than reason. As an example, he cited various aspects of the creeds of the Christian church, for which there could be no adequate rational proof.[45]

During the interwar period, however, Russell's primary concern with the Church was reflected not in an assault on its creeds but in how Christianity might relate to the formulation of a modern social ethic. What troubled Russell most was the relationship between individual and collective beliefs. Scornful in his rejection of dogma, he was still confronted with the need for a social ethic, and not merely a personal one, if society was to survive the dilemma presented by "the old savage in the new civilization." Whatever his distaste for creeds, therefore, he had to find some way around the problem that there could be no collectivity, in morality or in religion, without some degree of conformity. If there was any case to be made for a collective social ethic, there must be some form of universals that is the same among a range of individuals, so that their "personal moral codes" overlap to the extent that a social consensus is possible.

Russell was unable to accept that a "moral outlook" legitimized a religious or metaphysical perspective. This, in part, is why he is properly described not as a philosophical ethicist but as a social ethicist during the interwar period. If there must be some collective expression of personal morality in modern society, it could be expressed in terms of what needed to be done rather than in existential terms about meaning and truth. Thus he maintained that ethics was not properly a part of philosophy.[46] The difficulty, however, lay in how to effect the transition from personal to social ethics, and the basis on which any collective morality might be asserted. A purely individualistic ethic was useless for the old savage dilemma, because it was a dilemma faced by civilization as a whole. There needed to be a resolution in practical terms, whether or not there was any coherence in philosophical terms, and it had to take place within the context of scientific society in the Machine Age.

RELIGION, MORALITY, AND ETHICS

Russell could not accept the idea that morality in the modern world had to be derived from religion, much less that the history of Christianity should make it the most desirable religion from which to derive any such morality. He observed the disjuncture between the moral code expressed within the Gospel and what the Church, as a social institution, had come to espouse. Citing biblical injunctions to "turn the other cheek," "judge not lest ye be judged," and to give away one's wealth to the poor, he dryly pointed out that these maxims were excellent, "but not much practiced."[47] To charges that those without the Christian religion were wicked, he referred to the wickedness of those within the Church who committed the atrocities of the Inquisition, and so on. He then went on to discuss aspects of the gospel accounts of Jesus Christ, which reflected "defects" in Christ's character that were accepted as reasonable by the Church.[48]

Yet if the problem of the "old savage and the new civilization" was at heart a moral one, in that moral capacity had not kept pace with technological development, then there needed to be some basis on which to establish a scientific morality appropriate to the new demands that science and technology had placed on the individual in modern society. Whether morality of any kind was possible without a religious basis was another question; in one of his contributions to the *Today and Tomorrow* series, C. E. M. Joad struck a similar chord in identifying the prospects of morality in scientific society. Noting that "there has never been less" of a positive morality that might persuade people to pursue good things for their own sake than at the present time, Joad identified this lack with the decline of religion:

> A world without religion is a sad and tiring world because it lacks an object, and for this reason there have been few generations which have known less happiness than our own.[49]

Observing that the "emotional enthusiasm" that accompanies religion was essential to promoting this "true morality," Joad said the hope for the world lay in the revival of some kind of religion that might bring about this positive morality. Without religion, Joad implied, the only basis for morality would be what Thrasymachus advocated in Book I of Plato's *Republic:* the use of force, or the exercise of power.[51]

What I Believe (1925) contained the most succinct statement of how Russell thought to construct a basis for morality independent of religion. His second contribution to the *Today and Tomorrow* series contained "what I think of man's place in the universe and of his possibilities in the way of achieving the good life."[52] Whereas *Icarus* expressed his fears, *What I Believe* was supposed to express his hopes. Despite this good intention, Russell was roundly criticized for his "eclectic pessimism" by the Bishop of Ripon,[53] while Reinhold Niebuhr responded to Russell's hopes of "the good life" in an article entitled "Can Schweitzer save us from Russell?"[54] The majority of his discussion was about the nature of current morality and how a "scientific morality" might lead to a better chance of humanity achieving the good life. Russell began by placing man in the natural world, and then he dismissed the (finite) world of nature as being ultimately uninteresting, once all of the facts about it were known. Religious dogma had no place in this epistemological schema, for "God and immortality . . . find no support in science."[55] Summarizing the position he articulated in other works, he rejected the ontological status of electrons, protons, and the human soul as "logical fictions; each is really a history, a series of events, not a single persistent entity," and he called the antithesis of mind and matter "more or less illusory."[56] Once again, he identified fear as the basis of religious dogma, and he observed that religion allows us a share in omnipotence, when otherwise there would be physical limits to our control of the universe around us.[57] Thus, he said, a belief in God "humanizes" the otherwise indifferent world of nature.

This religious understanding of nature, which he rejected, came out of what he said was the confusion of a philosophy of nature with a philosophy of value. Cosmic anthropocentrism for Russell was an absurdity: "All such philosophies spring from self-importance, and are best corrected by a little astronomy."[58] There was no external source of value, for human beings alone conferred value on nature, for "outside human desires there is no moral standard."[59] Consequently, the good life, for Russell, was "one inspired by love and guided by knowledge,"[60] where love was the combination of "delight and well-wishing,"[61] and knowledge was not ethical knowledge but scientific knowledge and the knowledge of particular facts.[62] Russell was a consequentialist, dismissing the possibility of ethical knowledge, because he said that the decision about whether conduct was right or wrong could only be made with reference to its consequences: "Given an end to be achieved, it is a question for science to discover how to achieve it."[63] Therefore, he maintained that what distinguished ethics from science was not a special kind of knowledge but the desire to achieve a certain end.

Russell could not accept Jeremy Bentham's idea of enlightened self-interest as the path to happiness, however, saying that immediate satisfaction did not necessarily mean that in the long run individuals would act "rightly."[64] An individual conception of welfare had to be replaced by a social one in the modern world, he said.[65] The scientific moralist was to combat fear by increasing security and cultivating courage, though Russell warned there was "no short-cut to the good life, whether individual or social." "To build up the good life," he said, "we must build up intelligence, self-control, and sympathy."[66] In conclusion, he noted that nature, even human nature, would increasingly be what science manipulated it into being. Our freedom would only be won, he said, when we have the same control over our passions that we already have over "the physical forces of the external world."[67]

The separation Russell asserted between fact and value, and thus between knowledge and ethics, could explain why he did not develop the theory of impulse he had earlier outlined in *Principles of Social Reconstruction*. While *Principles* lacked the sophistication of his later efforts to understand the springs of human action, and how they might be channeled into creating a society less vulnerable to the adverse effects of science and technology, it was still an important attempt to understand the nonrational elements of human nature.[68] *Principles* dealt with what he later came to dismiss, in the interwar period, as something outside of the realm of knowledge. Instinct, impulse, emotion, and revelation, along with metaphysics as a whole, were not susceptible to the rational analysis he associated with "the scientific temper," and thus could not lead even to truth of a provisional nature. To solve the dilemma of "the old savage in the new civilization," and to prevent the start of the Next War, however, a social ethic had to consider and control both rational and nonrational desires, as well as the beliefs from which they sprang.

MYSTICISM AND PERSONAL MORALITY

It is significant that the only aspect of religion of which Russell wrote approvingly in the interwar period was that which involved a personal moral code.[69] Personal religious beliefs were unobjectionable, even in an age of science, as long as some kind of dogmatic hegemony was not pursued or advocated over the beliefs of other individuals. Religious belief thus became a matter of individual preference, and religion was seen to have value only as personal experience. Emotional experience, instinct, intuition, or revelation all might have their place in the religious beliefs of individuals, but each was incapable of leading to any social conclusions regarding knowledge of the physical world or any collective sense of the meaning of existence.

Russell's earlier attitude toward mysticism accommodated these ideas, in his insistence that mysticism was experiential, not epistemological, in character. Russell began his article on "Mysticism and Logic" for the *Hibbert Journal* in 1914 with a definition of metaphysics as having been derived "by the union and conflict of two very different human impulses, the one urging them towards mysticism, and the other urging them towards science."[70] Great philosophers, he said, had felt the need of both impulses, rendering philosophy "to some minds, a greater thing than either science or religion."[71] After using Heraclitus and Plato as examples of such great philosophers, Russell then went on to elucidate what he understood by mysticism.

First, mysticism was characterized by "the belief in insight as against discursive analytic knowledge."[72] The "definite beliefs" of mystics are the result of reflection "upon the inarticulate experience gained in the moment of insight."[73] There was, consequently, "the conception of a Reality behind the world of appearance and utterly different from it" that was revealed in this moment of insight.[74] The second fundamental aspect of mysticism, Russell maintained, was "the belief in unity, and [the] refusal to admit opposition or division anywhere," while the third was "the denial of the reality of Time."[75] Finally, he discussed the illusory nature of evil as something without a reality in itself but produced through "the divisions and oppositions of the analytic intellect."[76] He used these four "doctrines" to set up four questions regarding the truth or falsity of mysticism: first, the epistemological question about whether there are two ways of knowing (which he called "reason" and "intuition"); second, whether plurality or division was illusory or not; third, whether time was "unreal"; and fourth, whether there was some kind of reality to which "good" and "evil" belong.[77]

Russell dismantled each of the four doctrines he ascribed to mysticism but allowed for "an element of wisdom to be learned from the mystical way of feeling, which does not seem to be attainable in any other manner."[78] Consequently, he said, if he was correct, then mysticism was "to be commended as an attitude towards life, not a creed about the world."[79] Even the scientific spirit, he said, "may be fostered and nourished by that very spirit of reverence in which mysticism lives and moves."[80] It was a product of emotion, not of intellect, however, something that reflected emotional experience rather than knowledge. Any mystical creed, therefore, was "a mistaken outcome of the emotion," even though it might be "the inspirer of whatever is best in Man."[81]

Russell maintained that the "scientific attitude" required insight to be considered "an insufficient guarantee of truth, in spite of the fact that much of the most important truth is first suggested by its means."[82] He pointed to the common opposition that was seen between intuition and reason, and how, since the eighteenth century, each in turn had been thought to be the most important. He criticized Henri Bergson in this regard, for he said that Bergson "had raised instinct to the position of sole arbiter of metaphysical truth."[83] Russell, however, claimed the opposition between them was "illusory," saying that while instinct might lead to the beliefs that reason subsequently confirmed, the confirmation consisted in their agreement with other beliefs. While instinct was a creative force, "even in the most logical realm," reason was instead "a harmonizing, controlling force," one that examined the possibility of error.[84] There was thus no opposition between reason and intuition, except where some one-sided, "blind reliance" on instinct was required. In "the sphere of knowledge," Russell concluded, "scientific restraint and balance" was superior to "the self-assertion of a confident reliance on intuition." The same perspective may be applied to the epistemological problem of thinking that revelation was a means to the acquisition of truth. Russell could not accept that true knowledge might be acquired in any other fashion than through rational inquiry. If someone claimed such knowledge through revelation, moreover, there was no means of proving or disproving it other than by reason. He did, however, urge the importance of "that largeness of contemplation, that impersonal disinterestedness, and that freedom from practical preoccupations" he found in the mystics of "all the great religions of the world."[85]

In retrospect, we recognize the problems Russell himself would come to see in this representation of the relationship between knowledge, ethics, and experience. The disinterested stance of the philosopher became more difficult in the aftermath of the Great War. While he advocated a consideration of "the facts" alone, Russell himself tended to claim a universal character for his own assertion of "the facts" that others would dispute. The separation of logic, and thus knowledge, from mysticism (and emotional experience) also would continue to haunt Russell's attempt to formulate a moral outlook during the interwar period. Although he had dismissed evolutionism as an inferior form of "scientific philosophy," because of "its slavery to time, its ethical preoccupations, and its predominant interest in our mundane concerns and destiny,"[86] much of what he wrote in the interwar period could be similarly described, as he considered the practical answers required by the dilemma of "the old savage in the new civilization."

Before the War, therefore, Russell believed that mysticism provided individuals with an emotional experience that had personal validity, but it could not provide knowledge, and therefore could not be a source of truth, something that remained the preserve of science. During the interwar period, while he continued to deny that mysticism could be a source of knowledge and truth, Russell's description of a moral outlook assumed the characteristics of an "attitude toward life" on the part of both individuals and society as a whole. While mysticism could not provide knowledge of something other than what science could demonstrate, mystical experience itself had an emotional validity that Russell did not deny, even if it did not reveal anything that was absolutely "true" about reality. The pragmatic decisions required for the survival of the old savage, however, needed to be rational, based upon knowledge, and thus upon "truth," however provisional that truth might be. Unwilling to accept the utilitarian and mechanistic conception of society generated by the perception of science-as-technique alone, as we have seen, he was nudged toward explaining what the social expression of such an individual moral outlook might be. Yet by separating knowledge from experience in this way, Russell was excluding the nonmaterial aspects of human existence out of which he needed to construct his "moral outlook" against the wielders of scientific technique.

Despite Russell's aversion to the mechanistic outlook and his rejection of science-as-technique as a sufficient expression of what was meant by "science" in the modern world, the alternative "moral outlook" he suggested in *Prospects* and *The Scientific Outlook* remained not only indistinct but under siege from the outset. To uphold science-as-knowledge, Russell had to deal with the problem that he felt most knowledge was now negative knowledge, what could not be known about the universe, a very unsatisfactory conclusion for the Anglo-American public, which was entranced by the discoveries of relativity theory. Further, because of the demise of the materialist understanding of the cosmos, and his own recognition that entities such as atoms were "logical fictions," science-as-knowledge inevitably incorporated metaphysical elements. Once metaphysics gained a foothold in the scientific representation of nature, it would be difficult to deny the claims that apologists for religion had always made about the spiritual nature of the universe.

Russell therefore tried to exclude mysticism from science by claiming that it only had validity as an expression of personal, emotional experience, setting aside the problem that, for the individual, the acquisition of all knowledge is inseparable from experience of some kind. Further, if all truth discoverable by humanity was only provisional, then the knowledge acquired through science—the only means Russell would accept—was equally a provisional expression of truth, making dogmatic claims on behalf of science as invalid as dogmatic claims on behalf of any religion. Russell's moral outlook required something to offset the advantages that technique possessed in the organized, scientific society. An obvious source would have been religion, which expressed values other than those understood in terms of power and utility, yet Russell did not see how this could be possible, for several reasons. First, he had become completely disillusioned with the social expression of religion as a result of the Great War. The Christian church was an institution equally as archaic as any other in the Machine Age. Whatever its value for personal morality, therefore, religion could not be a source of social morality. Second, religion as a social phenomenon required the existence of creeds, to which a group of believers could assent. This involved an acceptance of dogma that Russell believed was inimical to the scientific spirit, in which skepticism was a necessity. The third and final blow to Russell's moral outlook, as we will see in Chapters 4 and 5, came with his acquiescence to the essential conflict between the dogmatic expression of revealed truth he associated with religion and the provisional truths discoverable through science. In the scientific society, he believed, religion and science could not be reconciled. Had he been as skeptical of the elements of the "public conversation" that asserted the inevitability of such a conflict as he was of the metaphysical speculations about a reconciliation between science and religion, however, his own conclusions might not have been so dogmatic.

Chapter 4

Physics and Philosophy

After the Great War, science and technology prompted more than the question, "What ought we to do?" The discoveries in physics and astronomy that followed Albert Einstein's papers on special and general relativity (and the demonstration of some of his ideas during the solar eclipse of 1919) also prompted the question, "What does it mean?" With the fundamental axioms of the so-called Newtonian universe in pieces, relativity theory and quantum mechanics challenged the previous understanding of the physical universe and humanity's place within it. In physics, cosmology and cosmogony were intertwined; the nature of the physical universe became inseparable from theories about its origin and its ultimate end. Quite apart from the moral questions that emerged from the Great War, science itself prompted a questioning of the nature of reality during the interwar period in a way that invited a metaphysical explanation or a religious response.

Russell complained in the 1914 edition of *Our Knowledge of the External World* that physicists had not explored the philosophical implications of their work.[1] He was one of the people who attempted to make philosophical sense of recent discoveries in the physical sciences during the interwar period, with the publication of *The ABC of Relativity, The ABC of Atoms, The Analysis of Mind,* and *The Analysis of Matter,* as well as a number of articles and reviews on similar topics. By 1927, Russell included in *An Outline of Philosophy* a concise account of the salient features of the new physics and their philosophical implications. That same year, Arthur Eddington published his Gifford Lectures under the title, *The Nature of the Physical World,* and he undoubtedly made Russell wish that physicists would leave philosophy alone.

Although Russell made deferential use of Eddington's earlier works, *Space, Time and Gravitation* (1921) and *The Mathematical Theory of Relativity* (1925), when physics merged into metaphysics, the Quaker physicist was unwilling to defer in a similar fashion to the limits Russell placed on what we can know about the universe. *The Nature of the Physical World* (1928) was written in an accessible style, and it enjoyed considerable popularity, and some notoriety, for Eddington's speculations about what the new physics might mean to an understanding of what lay beyond the physical world. At the same time, new theories of the origins of the cosmos emerged from discoveries in astronomy, making cosmogony, cosmology, and universal entropy topics of popular as well as academic discussion. The public was presented with the novelty of reputable scientists such as Eddington and James Jeans musing in a very unscientific fashion on the origin, meaning, and ultimate ends of life.

After the success of his "shilling shocker," *The Problems of Philosophy* (1912), Russell was aware of the enormous market for popularly written books and the corresponding influence that such books might have on public opinion. The opportunities in the interwar period for the mass dissemination of ideas encouraged other kinds of propaganda than the strictly political, and Russell understood how the scientific society could be manipulated in various directions as a result. His concern with "the philosophical implications of the new physics" was twofold: first, he believed it was important to clarify just what the new physics revealed; second, it was essential to clarify what the new physics implied about the questions of meaning and purpose on which scientists such as Eddington and Jeans chose to speculate. In effect, he was less concerned about the arguments themselves than their popularity and the likelihood that they would be believed. In Russell's version of science in the interwar period, Jeans and Eddington were made representative of scientists as a whole, their fallacies typical of the fallacious cast of current trends in science. J. B. S. Haldane, in a review of *The Scientific Outlook,* however, commented that "Russell contrives to knock the heads of his distinguished colleagues together with a resounding crack," but had not demonstrated that "the bulk of eminent physicists have made pronouncements that materialism is disproved and religion reestablished."[2] Thus it is more accurate to describe Russell's exchange with Eddington and Jeans as an argument in the popular press, the product of different perceptions of cosmology contending for the attention of a popular rather than an academic or a scientific audience. "Rhetorical performance," therefore, was an important element of what he wrote about these two physicists, even more so than it was in his earlier popular accounts of the meaning of the new physics. To understand the nature of this performance, we must first consider what Russell thought to be the philosophical implications of the new physics and then identify what disturbed him so deeply about the speculations of Eddington and Jeans.

THE NEW PHYSICS

The philosophical implications of the new physics stemmed in part from the fact that the physical world revealed by relativity theory and quantum mechanics was nothing like the common-sense "reality" of the world of everyday life.[3] In effect, the new physics was counter-intuitive, at least as far as the world depicted by Newtonian mechanics and classical materialism was concerned. "Matter" was revealed to be a metaphysical concept; without the definition of substance that went with "matter," the understanding of causality required a radical revision. Then came the question of time: with relativity theory, the notion of a single cosmic time, in which chronological relationships could be discussed, was replaced by a multiplicity of times dependent only on the frame of reference in which they were measured. Doubt also was cast on what was revealed through perception; the role played by physiology in the human experience of the physical world related mind and matter in a new way. This was the result of what Russell called the demise of "scientific materialism of the eighteenth-century variety."[4]

Russell dealt with some of the implications of the new physics in *An Outline of Philosophy* and the publications that preceded it. Not surprisingly, he was most interested in what mathematical approaches produced in the way of knowledge of the physical world; any knowledge that was not mathematical, he believed, rested on a much more shaky scientific base. Thus, for Russell, the new physics produced more negative than positive knowledge about the physical world than had classical physics: "Physics is mathematical, not because we know so much about the physical world, but because we know so little: it is only its mathematical properties that we can discover. For the rest, our knowledge is negative."[5] This distinction between "negative" and "positive" knowledge lay at the heart of Russell's epistemology, at least insofar as it applied to his view of the philosophical implications of the new physics. Russell prided himself on being a skeptic, on not accepting as true (even provisionally) something that could not adequately be demonstrated.[6] Given the stringent terms that he set for such a demonstration (in the work of other people, if not in his own), it is not surprising that his dismissal of "metaphysical speculations" frequently rested upon the failure of their expositors to meet his requirements.[7] His efforts were thus directed toward exposing untenable propositions, whether in science, religion, or philosophy. This meant that he dealt more with what could not be known than with what could. As his retreat from Pythagoras progressed, though, Russell was forced to advance propositions of his own about the nature of reality, and about the kind of social ethic required in the "new civilization." It was insufficient to state what could *not* be known; immediate decisions had to be taken concerning the future of civilization. For a truly scientific society to emerge from those decisions, some kind of positive knowledge, however provisional, was required.

THE DEMISE OF MATTER

Russell was quite emphatic about the philosophical significance of the changes wrought by quantum mechanics in the conception of "matter." As far as he was concerned, "the main point for the philosopher in the modern theory was the disappearance of matter as a 'thing.'"[8] This metaphysical shift away from a strictly material definition of substance, he said, changed the very subject of physics from the study of "bodies in motion" to the study of "events."[9] Whereas classical mechanics understood there to be some ultimate particle, the doctrine of atomism had been superseded by a quantum model of atomic structure, which by 1927 had cast more than a little doubt on such a certainty. The interior of the atom was now understood to be comprised of orbits of electrons, around a nucleus of protons and neutrons; physicists realized that the only way they knew of an electron's existence was when it moved, or jumped, from one quantum level of energy to another. Thus the existence and character of a subatomic "event" could be measured but not the "thing" that participated in such an event:

> Modern physics, therefore, reduces matter to a set of events which proceed outward from a centre. If there is something further in the centre itself, we cannot know about it, and it is irrelevant to physics. The events that take the place of matter in the old sense are inferred from their effect on eyes, photographic plates, and other instruments. What we know about them is not their intrinsic character, but their structure and their mathematical laws.[10]

Although the word "matter" still served as a convenient shorthand for describing the physical world as it appeared to common sense, in philosophical terms it had ceased for Russell and his contemporaries in the new physics to have anything like its previous meaning. As Russell explained in *The ABC of Relativity*, relativity "demands the abandonment of the old conception of 'matter,' which is infected by the metaphysics associated with 'substance.'"[11]

Thus "the electron ceases altogether to have the properties of a 'thing' as conceived by common sense; it is merely a region from which energy may radiate."[12] As for what we may know about these regions, since we have no evidence about what goes on when an atom is not absorbing or radiating energy, Russell pointed out that "consequently all evidence as to atoms is as to their changes, not as to their steady states."[13] To be concerned with changes is to require an understanding of time; when coupled with the demise of "scientific materialism," the depiction of time in relativity theory conspired to make our knowledge of the physical world even more tenuous.[14]

Instead of a world of space and time, in which bodies occupy a specific space for a period of time, the new physics presented instead the world of space/time, in which the coordinates of any body needed to include the fourth dimension of time, as well as the three dimensions of classical "space." Thus, as Russell put it in

The ABC of Atoms, "the ultimate facts in physics must be events, rather than bodies in motion."[15] Further, the relationship was established between matter and energy, to the point it made no sense to speak of the "essential" quality of matter without referring to the energy states of its particles, and both needed to be expressed in terms of duration. Just as quantum theory disposed of the idea of a persistent "space," however, relativity theory disposed of the idea of "one cosmic time."[16] What was more, the commonsense view of time held there was a definite answer to the chronological order of specific events; relativity showed this perception to be wrong.[17]

If quantum theory recast the understanding of matter, relativity theory recast the understanding of motion. Einstein's initial premise was that motion is relative to a frame of reference; the motion of any body is never absolute but instead is relative to the frame of reference of the observer. The complete separation of space and time he identified with "the old view" meant that "a piece of matter was something which survived all through time, while never being at more than one place at a given time."[18] Pieces of matter are thus replaced in space/time by "events," which do not "persist and move" but which exist for "a little moment and then cease to exist."[19] Consequently, as these particles are extended in time, "the whole series of these events makes up the whole history of the particle, and the particle is regarded as *being* its history, not some metaphysical entity to which the events happen."[20] The metaphysical status of the particle is further eroded by the conclusion it is only known by its "effects"; if this is the case, then as Russell says, "there is no reason to suppose anything exists except the effects."[21]

In summary, the space and time of classical mechanics gave way to the "space-time" of relativity theory and quantum mechanics. Just as the objective status of a material body was no longer a tenable idea, so too the objective measure of time had been supplanted by one dependent solely on the frame of reference in which it was measured. Matter was rendered into a series of events, and time became relative instead of absolute; this necessitated a rethinking of what classical mechanics meant by "causality."

The cause-effect relationships of the commonsense world were replaced by a succession of events in space-time, and this succession of events in turn became definitive, at least of the structure of matter, if not of its intrinsic nature.[22] If there is no such entity as a "body," then Newtonian concepts of "force" become problematic; instead of a universe where "every action (on a body always) has an equal and opposite reaction," there is a universe in which such absolute statements have no place, because "what we have to substitute for force is laws of correlation."[23] This is not to say that the commonsense view of causality did not have its uses; it was just that the relativistic universe necessitated a less dogmatic approach to what constituted a natural "law":

> At an early stage of a science this point of view is useful; it gives laws which are usually true, though probably not always, and it affords the basis for more exact laws. But it has no philosophical validity, and is superseded in science as soon as we arrive at genuine laws. Genuine laws, in advanced sciences, are practically always quantitative laws of *tendency*.[24]

Sidestepping the difficulty that results from describing natural *laws* as mere "tendencies," he next dealt with how we are forced to ask questions about the nature of perception if we are to deny the physical reality of what we perceive through our senses in favor of the mathematical structure of a relativistic universe.

Russell bluntly asserted the existence of a metaphysical gulf between the physical and the perceptual universe:

> Physical space is neutral and public: in this space, all my percepts are in my head, even the most distant star is *as I see it*. Physical and perceptual space have relations, but they are not identical, and failure to grasp the difference between them is a potent source of confusion.[25]

Further, because all of our knowledge of the physical world "must start from percepts," at least in the beginning this gives "a subjective cast to the philosophy of physics."[26] This subjectivity is made more extreme, Russell contended, because "there is no direct spatial relation between what one person sees and what another sees, because no two persons ever see exactly the same object." As a result, "this shows how entirely physical space is a matter of inference and construction."[27] The result of such a construction is the acceptance of the independence of perception and the object perceived:

Thus what is called a perception is only connected with its object through the laws of physics. Its relation to the object is causal and mathematical; we cannot say whether or not it resembles the object in any intrinsic respect, except that both it and the object are brief events in space-time.[28]

To erode even further any confidence in the knowledge of the physical world gained through perception, Russell whimsically remarked that there also was no certain evidence that the world revealed by perception is not in fact what is "really there."[29] This lack of absolute knowledge of the intrinsic nature of the physical world contributed, in its turn, to the resolution of the traditional duality between mind and matter: "The stuff of the world may be called physical or mental or both or neither, as we please; in fact, the words serve no purpose."[30]

Because the new physics denied any verifiable knowledge of the intrinsic character of matter, this in large measure was the reason for collapsing the so-called Cartesian duality between mind and matter. Russell believed that this dualism was "mistaken," and he asserted that if "metaphysics is ever to be got straight," the confusion over mind and matter had to be resolved.[31] In fact, he said, "the gap between mind and matter has been filled in, partly by new views on mind, but much more by the realization that physics tells us nothing as to the intrinsic character of matter."[32] In *The Analysis of Mind,* Russell concluded that mind, as well as matter, was a "logical construction,"[33] and he proposed a fundamental relation between the two in terms of a "neutral stuff," out of which both were derived.[34] By the time *An Outline of Philosophy* appeared, Russell had refined his argument about what he called "neutral monism":

> It is monism in the sense that it regards the world as composed of only one kind of stuff, namely events; but it is pluralism in the sense that it admits the existence of a great multiplicity of events, each minimal event being a logically self-subsistent entity.[35]

Whether or not this concept actually *resolved* the mind/matter duality, it provided a kind of closure beyond merely informing his readers what they could *not* know about the physical world.

Yet this was precisely Russell's problem. People wanted to know something about the world or about the cosmos; there is something unsettling about being stripped of familiar ideas about mind and matter, for example, without being given a replacement. It is arguable that even Russell himself was not content with merely "negative knowledge." After summing up the negative knowledge provided by the new physics, he concluded his *Outline* with "a few words about man's place in the universe."[36] Because the vast universe revealed by recent discoveries in astronomy was neither friendly nor hostile, he said humanity should face it in the mood of "quiet self-respect," which philosophy made possible.[37] "Philosophy," he declaimed, "comes as near as possible for human beings to that large, impartial contemplation of the universe as a whole which raises us for the moment above our purely personal destiny."[38] He thought the philosophical perspective suggested (in the sense of positive knowledge) that the physical world was "perhaps less rigidly determined by causal laws than it was thought to be," to the point "one might . . . attribute even to the atom a kind of limited free will." Thus, he said, "There is no need to think of ourselves as powerless and small in the midst of vast cosmic forces."

Russell's genial attitude in *The Outline of Philosophy* toward the possibility of free will for atoms underwent a dramatic change by the time *Religion and Science* was published. The reason for this changed attitude was undoubtedly the popularity of Sir Arthur Eddington's view on the philosophical implications of the new physics in *The Nature of the Physical World*.[39] Suggestions might be made about "man's place in the universe," but when they appeared in the context of a best-selling book by a prominent scientist, Russell reassumed the guise of the complete skeptic, for whom most of the knowledge revealed by physics was negative knowledge, or what could not be known, about the nature of the physical world.

SCIENCE AS KNOWLEDGE

Russell's operating definition of "scientific method" was simple enough to allow a retrojection of "science" into distant periods of human history. In essence, he said, it consisted of observing facts that enabled the observer to discover general laws that in turn governed those specific facts. The two stages of observation and inference were essential and "susceptible of almost indefinite refinement."[40] Yet he wryly observed how unnatural the scientific method was for human beings outside of the laboratory, even for scientists themselves. Picking up something of Freud's discussion of human psychology, Russell asserted that the majority of human opinions are "wish-fulfillments, like dreams in Freudian theory."[41] Even the minds "of the most rational among us," he said, "may be compared to a stormy ocean of passionate convictions based upon desire, upon which float perilously a few tiny boats carrying a cargo of scientifically tested beliefs."[42] This was not altogether a bad thing, nor should it be unexpected, Russell maintained. No one could live for long without "a certain wholesome rashness," for there was seldom time to test rationally all of the opinions that underlie human conduct.[43] He considered a scientific opinion to be "one which there is some reason to believe is true," while an unscientific opinion was one "which is held for some other reason than its probable truth."[44] This, of course, made the nature of "truth" central to Russell's depiction of science and the perception of truth integral to the scientific outlook.

For all his concern with the scientific outlook, however, Russell's characterization of scientific method was at best perfunctory and made little contribution to the philosophy of science. Scientific knowledge was, for him, part of an inductive hierarchy, moving from "instances" to hypotheses, and then to increasingly more general laws. His depiction of the process by which this hierarchy was established was rudimentary. Starting with the observation of facts, it led to a hypothesis, and then to the testing of that hypothesis by further observation.[45] As his example of this inductive method, he cited physics as the science that had come closest to this model of "perfection." Having established scientific method as an inductive hierarchy, Russell then undercut its claims to absolute truth by insisting that science proceeded by approximation,[46] and that induction was logically no means to achieve certainty: "All scientific laws rest upon induction, which, considered as a logical process, is open to doubt, and not capable of giving certainty."[47] Having said this, Russell still defended science as a process of successive approximations toward exact truth, and he went further by saying mathematical technique was not the only means of arriving at that truth. He cited Pavlov's work as an example of scientific method, although it did not proceed along lines conducive to mathematical analysis.[48]

Not only did Russell fail to provide a significant account of how the scientific method might be applied to problems in science, he also detailed its limitations in a way guaranteed to promote a healthy skepticism toward the more extravagant claims scientists might make. He asserted the importance of a verification principle for all scientific knowledge, whether or not that knowledge still had to be accepted on authority by most individuals. Consequently, one of the features of the advance of science was that less and less of what was known turned out to be "datum" or knowledge, and more and more was found instead to be "inference."[49] He then represented the limitations of the scientific method under three headings: the doubt about the validity of induction; the difficulty of drawing inferences from what is experienced to what is not experienced; and, if such a possibility exists, the problem that the extrapolated inference "must be of an extremely abstract character" giving "less information than it appears to do when ordinary language is employed."[50]

To support the first limitation, Russell synopsized Hume's critique of induction and dispatched Hume's critics by saying: "It is easy to invent a metaphysic which will have as a consequence that induction is valid, and many men have done so; but they have not shown any reason to believe in their metaphysic except that it was pleasant."[51] Thus, he said, although there may be "valid grounds" for believing in induction (and "none of us can help believing in it"), it remained "an unsolved problem of logic," albeit one with appropriate safeguards, which might still be used.[52] As for the problem of inference, Russell observed that little of what is known can be experienced directly; much of this ostensible "knowledge from experience" was, in fact, knowledge gained through inference. The conclusion physicists had reached was to return to a kind of Berkeleyan idealism in which the world of matter ceased to exist: "That in itself, however, would be no great loss, provided we could still have a large and varied external world, but unfortunately they have not supplied us with any reason for believing in a non-material external world."[53] Saying that this problem also was one for the logicians, he once again avoided an answer of his own, commenting that until this problem of inference was resolved, "our faith in the external world must be merely an animal faith," the faith to which physicists had gradually "turned traitor":

> The fact is that science started with a large amount of what Santayana calls "animal faith," which is, in fact, thought dominated by the principle of the conditioned reflex. It was this animal faith that enabled physicists to believe in a world of matter.[54]

The problem that physicists believed in a world of matter only because of their "animal faith" was compounded by the third limitation of the scientific method, which in his illustration Russell called "the abstractness of physics." Even if we are able to extrapolate our inferences, he said, it must be in the form of a mathematical abstraction appropriate to the abstract character of that inference: "Ordinary language is totally unsuited for expressing what physics really asserts, since the words of everyday life are not sufficiently abstract. Only mathematics and mathematical logic can say as little as the physicist means to say."[55] Unlike his discussion of the other two limitations, here the purpose behind Russell's discussion of scientific method became more apparent. What disturbed him the most about modern physics, it seemed, was not the science itself but what physicists had tried to communicate to the public concerning their inferences about the nature of the cosmos and humanity's place within it:

> As soon as [the physicist] translates his symbols into words, he inevitably says something much too concrete, and gives his readers a cheerful impression of something imaginable and intelligible, which is much more pleasant and everyday than what he is trying to convey.[56]

He identified this impulse on the part of many people with "a passionate hatred of abstraction," which he attributed to "its intellectual difficulty." Yet scientific thought, by its very nature, he considered abstract, because it dealt with causal laws. It was "essentially power-thought," whose purpose was to give some kind of power, conscious or unconscious, to the possessor. "The more irrelevant details we can omit from our purview," he said, "the more powerful our thoughts will become." He cited the difference between the money a farmer makes from his crops and that made by someone who manipulates the stock market (who, "of all those concerned in the economic sphere, makes the most money and has the most power").[57] Russell concluded that, due to its "extreme abstractness," modern physics alone was able to supply those who understood it with "a grasp of the world as a whole." Thus, he said, "the power of using abstractions is the essence of intellect, and with every increase in abstraction the intellectual triumphs of science are enhanced."[58]

Recent discoveries in physics had made some physicists tentative about what they knew about the cosmos. Russell called this result a loss of faith, "a curious fact that, just when the man in [the] street has begun to believe thoroughly in science, the man in the laboratory has begun to lose his faith."[59] The revolutionary ideas of the philosophy of physics, he said, have come from the physicists, making the new philosophy "humble and stammering" where the old one was "proud and dictatorial."[60] Into the vacuum left by the disappearance of belief in physical laws have tumbled "any odds and ends of unfounded belief," and Russell bitterly observed, "we must expect the decay of the scientific faith to lead to a recrudescence of pre-scientific superstitions."[61] His prime example of this decay was found in Eddington's *The Nature of the Physical World*.

SCIENCE AND THE UNSEEN WORLD

Arthur Eddington's publications prior to 1927 were treated respectfully by Russell.[62] Certainly, there was much even in *The Nature of the Physical World* that Russell found unobjectionable. In most of the book, Eddington discussed the developments of the new physics in a fashion similar to Russell's.[63] The final four chapters, however, dealt with the philosophical implications of the new physics in a different light. Eddington's central observation was on the inadequacy of the materialistic analysis inherent in nineteenth-century classical physics. He pointed to Heisenberg's Indeterminacy Principle (where it was demonstrated that we can know *either* the position *or* the speed of a particle, but not both) as evidence that materialism is insufficient to describe the universe as it *is*. As a result, he claimed that "scientific investigation does not lead to knowledge of the intrinsic nature of things,"[64] and consequently he implied the overthrow of "strict causality."

As a consequence of Eddington's division of the domain of experience into the scientific and the extra-scientific, between the metrical and the non-metrical, however, he admitted the likelihood of there being more knowledge of the physical world than strictly mathematical knowledge of its structure. Whereas Russell was content with the "negative knowledge" physics supplied about the physical world, Eddington was not. The world of "pointer readings" was inadequate to explain the totality of the universe, he said, because a wholly physical interpretation includes "an hiatus in reasoning" by excluding what Eddington called the "spiritual world."[65] As evidence for this persuasion, he argued that mind was the underlying reality of experience.[66]

Though Eddington was obviously intrigued by this notion, he did not delve too deeply into its implications in his own description of the "physical world" in the first two-thirds of his book. Yet where Russell the philosopher was content with the tenuous character of the physical world revealed by quantum mechanics and relativity theory, Eddington the physicist was not content with the tenuous character of the philosophy it had generated. Whereas the overthrow of strict mechanical causality liberated Russell's universe from absolute determinism (at least in *The Outline of Philosophy*), it meant for Eddington that religion became possible for the scientific man.[67] If there were limits to the domain of physical science, this meant there was something beside the world of matter, that physics could neither prove nor disprove.[68] The existence of a non-metrical realm, whatever it might be called, was not dependent upon science. Thus, he said, "the religious reader may well be content that I have not offered him a God revealed by the quantum theory, and therefore liable to be swept away in the next scientific revolution."[69]

Of all the material in *The Nature of the Physical World*, comments such as these in the final few chapters seemed to have elicited the most heated response, so much so that two years later, Eddington's Swarthmore Lectures were published as a continuation.[70] In *Science and the Unseen World*, he developed his idea that the world represented by physics was "a symbolic world, inhabitable only by symbols."[71] Thus, Eddington said, "our environment may and should mean something towards us which is not to be measured with the tools of the physicist or described by the metrical symbols of the mathematician."[72] This "something more" was related to Eddington's belief that "Mind is the first and most direct thing in our experience."[73] Because the new physics led not to "concrete realities" but to "a shadow world of symbols," we return to the starting point in human consciousness looking for something more. It is at this point that reasoning fails us, he said:

> The premises for our reasoning about the visible universe, as well as for our reasoning about the unseen world, are in the self-knowledge of the mind. . . . Consciousness can alone determine the validity of its convictions.[74]

Thus, he concluded, "we have to build the spiritual world out of symbols taken from our own personality, as we build the scientific world out of the symbols of the mathematician."[75]

In a review of *The Nature of the Physical World*, entitled "Physics and Theology," which was published in *The Nation*, Russell responded negatively to Eddington's metaphysical speculations, although he had praise for the sections on the new physics.[76] It was the last quarter of the book with which he took issue, those chapters that were "of most interest to the general reader," and were "devoted to the exposition of an idealist philosophy and the advocacy of free will." Russell sharply repudiated Eddington's version of idealism by noting that, however much he knew about physics, "being no psychologist he exaggerates what psychology can tell us about the mental world." While Eddington believed direct self-knowledge was possible, Russell observed that such an opinion could "not survive a scrutiny of what is meant by knowledge." Similarly, he dismissed Eddington's "mind-stuff" almost in passing, "because I hold that mentality is a form of organization, not a property of individual events, just as, say, democracy is a property of a community and not of an individual citizen."

In *The Scientific Outlook*, Russell rejected Eddington's description of the laws of classical physics as nothing more than "conventions as to measurement" and his depiction of quantum mechanics and chance in a way that precluded the application of the law of causality to the actions of individual particles.[77] What exercised Russell the most about Eddington's book was his application to the cosmos of the Second Law of Thermodynamics, or the law of entropy, according to which Eddington concluded the universe was winding down to its inevitable end. Eddington's statement of the "principle of indeterminacy," moreover, that one can know only the position or velocity of a particle at a given time, but not both, disturbed Russell greatly because of how the physicist used it to eliminate physical determinism and to "rehabilitate" free will.[78]

It is ironic that Russell did not dispute the science behind Eddington's conclusions, so much as he assailed its philosophical implications. He fumed that "it speaks well for Sir Arthur's temperamental cheerfulness" that he should find in the inevitable demise of the universe "a basis for optimism,"[79] for if Eddington was right, this was virtually all physics could tell us, "since all the rest is merely rules of the game."[80] According to this view, evolution (as the increase of organization in one part of the universe) was futile, because the world would eventually be swallowed up in the general disorganization or entropy that would one day make the universe "a uniform mass at a uniform temperature."[81] Far from viewing this dispassionately as a scientific hypothesis, Russell complained that, "from a pragmatic or political point of view," such a theory would destroy, "if it becomes widespread, that faith in science which has been the only constructive creed of modern times, and the source of practically all change both for good and for evil."[82] Despite his earlier rejection of Albert Schweitzer's view, that ethics and morality needed to be grounded in a particular worldview, here Russell attributed the development of modern science to the belief in a rational universe governed by natural law. Lose that belief, he said, and replace it with skepticism, and it may lead "in the end to the collapse of the scientific era, just as the theological skepticism of the Renaissance has led gradually to the collapse of the theological era."[83] The crux of the problem for Russell was the question of what science could contribute to metaphysics. Eddington's theory by itself, as a scientific hypothesis, was one thing. As an interpretation of the ultimate end or meaning of existence, it was quite another.

As for the influence of mind on matter, Russell was not happy with Eddington's suggestion that the mind could somehow affect statistical probabilities. While both men agreed that the implication of quantum mechanics was the indeterminateness of the actions of individual atoms, Russell seemed to lean toward the idea that the statistical probabilities that predicted the behavior of large aggregates of atoms indicated physical determinism of a sort.

Eddington's view, however, was that the mind could interfere with "the laws of chance," and although Russell did not entirely discount the possibility, he said there was no evidence to support it either.[84] He noted, however, that Eddington did not seem to believe that mind could counter universal entropy, and thus he seemed "to accept with equanimity the view, which follows from this law, that the world is running down and must in the end cease to contain anything interesting." Saying it required "a certain robustness to be optimistic on the basis of such a philosophy," Russell summed up Eddington's theology by observing that "he has apparently a God, but his God seems to have made the world long ago and forgotten all about it."[85]

THE MYSTERIOUS UNIVERSE

In contrast to Eddington's God, the God envisioned by James Jeans was thinking about the world all of the time. The author of several books on astronomy and cosmology, Jeans published *The Mysterious Universe* in 1930, continuing the public discussion of the philosophical implications of the new physics that Eddington's book had initiated. In it, he picked up on the idea that mind somehow underlies the material universe when he mulled over the possibility that the universe is a world of pure thought.[86] This meant for him that creation itself was an act of thought, and presupposed some kind of "divinity" that did the thinking. Like Russell, he considered the dualism of mind and matter specious,[87] but instead he supported the idea of a universal mind or logos that underlay physical reality.[88] Of course, this necessitated the same turn away from a mechanistic understanding of nature as Eddington had advanced: "In a completely objective survey of the situation, the outstanding fact would seem to be that mechanics has already shot its bolt and has failed dismally, on both the scientific and philosophical side."[89]

Like Eddington, Jeans applauded the loss of the old determinism, and he proceeded from this to a conclusion about the universe quite unlike the ones reached by Russell:

> The picture of the universe presented by the new physics contains more room than did the old mechanical picture for life and con-

sciousness to exist within the picture itself. . . . For aught we know or for aught that the new science can say to the contrary, the gods which play the part of fate to the atoms of our brains may be our own minds.[90]

Like Russell, however, Jeans recognized that the idea of "substance" had been eradicated by the new physics; like Russell, he made the connection between physical reality and the "abstract creations" of the mathematician's mind. From this point, however, Jeans took his argument in an utterly different direction: if mind and matter are related, he said, we can postulate a universe of pure thought in which "matter" is one form in which thoughts may appear. The uniformity of nature itself argues for a self-consistency that can be seen as "the laws of thought of a universal mind."[91] Thus, he concluded, "If the universe is a universe of thought, then its creation must have been an act of thought. Indeed, the finiteness of time and space almost compel us, of themselves, to picture the creation as an act of thought."[92]

The prospect of some "universal Mind" thinking the universe into existence was obviously not one of the philosophical implications that Russell derived from the new physics. Given Jeans's stature in the scientific community, and the immense popularity of his book, he was too significant a figure for Russell, in his role as skeptic, to ignore. In a review of *The Mysterious Universe,* he disparaged Jeans's speculations even more curtly than he had Eddington's. Observing that the last quarter of the book "is occupied with matters on which the author does not speak as an expert," he brusquely dismissed Jeans's mathematical version of the Deity who ordered the universe:

> This part has pleased the theologians. It is nowadays expected of all eminent men of science that they should join in the defense of property against the Bolsheviks by showing that God made the world and therefore the capitalist system. Theologians have grown grateful for small mercies, and they do not much care what sort of God the man of science gives them so long as he gives them one at all.[93]

Jeans's idealism, in which the existence of material things is the product of their being thoughts in the mind of God, got equally short shrift from Russell. He also disputed Jeans's identification of mathematical symbols with the use that physics makes of mathematics to depict the physical universe.[94] Finally, Russell lampooned Jeans's hubris in personally attempting to solve the riddle of the universe, leaving no doubt regarding what he thought about scientists who strayed so far from their domain.[95]

Russell demonstrated a significant concern for the worldview that seemed to be emerging from the popular discussion of the cosmological implications of quantum physics. His reaction to Eddington and Jeans, or more specifically to *The Nature of the Physical World* and *The Mysterious Universe,* was part of his larger concern with the direction in which discussions of the relationship between science and religion had been moving since the end of the Great War. He was deeply distrustful of a facile reconciliation between science and religion, mostly because he feared what would happen if the problems of this world were explained or dismissed in metaphysical terms. In the speculations of Eddington, Jeans, and others, Russell saw no interesting possibilities, just bad philosophy, and bad philosophy could not yield any positive solution to the dilemmas faced by "the old savage in the new civilization."

Russell's predilection for skepticism when it came to the philosophical implications of physics, his preference for the certainty of "negative" knowledge over "positive" knowledge about the nature of the physical world of "facts," carried over into his discussion of the world of "values." If matter had ceased to be material, if so little could be known through the operations of science about the physical world, then any assertions of certainty in the world of values were inevitably invalid. Values might exist in the realm of personal experience, but he maintained they had no place in the realm of knowledge.

FAITH AND SKEPTICISM IN SCIENCE

The relationship between "fact" and "value," which may be paralleled to that between "knowledge" and "belief," was inseparable from the debate in the interwar period over the relationship between science and religion. The supposed conflict between a "scientific" outlook and a "religious" outlook, because of the extent to which it was debated at a popular level, led to speculations about the meaning and purpose of existence in the new civilization, speculations that disturbed Russell to a considerable extent. His own interpretation of the current impasse was significant in his attempts at articulating a social ethic for the scientific society. He called it a paradox, that just as fundamental science was replacing "the Newtonian order and solidity" with "a world of unreal and fantastic dreams," applied science was becoming "peculiarly useful."[96] He asserted that science played "two quite distinct roles: on the one hand as a metaphysic, and on the other hand as educated common sense."[97] As a metaphysic, science had been undermined by its own success, because "mathematical technique" could generate a formula for any conceivable world.[98]

In the last chapter of his section on scientific method in *The Scientific Outlook*, "Science and Religion," Russell rehearsed some of the arguments he would go on to make in *Religion and Science*. In dismissing Eddington, Jeans, and the "biological theologians," he said they agreed "that in the last resort science should abdicate before what is called the religious consciousness," something he considered "the outcome of discouragement and loss of faith."[99] Whatever his specific objections to Eddington's view of free will, or Jeans's depiction of a "mathematical God," or a divine purpose emerging through evolution, all of these metaphysical speculations foundered on the problems posed for society by the operation of science-as-power, or scientific technique:

> It is not by going backward that we shall find an issue from our troubles. No slothful relapses into infantile fantasies will direct the new power which men have derived from science into the right channels; nor will philosophic scepticism as to the foundations arrest the course of scientific technique in the world of affairs. Men need a faith which is robust and real, not timid and half-hearted.[100]

This relapse into "infantile fantasies" seemed to be the direction in which modern physics was heading. Russell complained that physicists were "so pained by this type of conclusion to which logic would have led them that they have been abandoning logic for theology in shoals," thereby trading reason for unreason and reality for unreality.[101] Russell could not comprehend why "religious apologists" would welcome the idea that there were no laws of nature, or that the natural order was due "to our own passion for pigeon holes."[102] The eighteenth-century idea of natural law seemed to be replaced by an irrational universe, in which these apologists had been made in God's irrational image.[103]

Having so pungently dismissed the metaphysical beliefs of Eddington and others, Russell then stated his own. Against the idea that the universe was a unity, which had been taken over from academic philosophers by contemporary "clergymen and journalists," he said:

> The most fundamental of my intellectual beliefs is that this is rubbish. I think the universe is all spots and jumps, without unity, without continuity, without coherence or orderliness of any other of the properties that governesses love. Indeed, there is little but prejudice and habit to be said for the view that there is a world at all.[104]

> I think that the external world may be an illusion, but if it exists, it consists of events, short, small, and haphazard. Order, unity and continuity are human inventions just as truly as are catalogues and encyclopedias.[105]

This did not affect his perception of the utility of science considered as common sense, however, because he believed that at a practical level, regardless of their metaphysical verity, scientific beliefs could have a material effect on the conduct of human affairs. Thus, he said, when he spoke of the importance of scientific method, he was referring not to its metaphysical status but to "its more mundane forms."

As far as Russell was concerned, science-as-metaphysics belonged with "religion and art and love, with the pursuit of the beatific vision, with the Promethean madness that leads the greatest men to strive to become gods."[106] After speculating that "perhaps the only ultimate value of human life" was to be found "in this Promethean madness," he proceeded to call this value a religious one, "not political, or even moral."[107] He said that it was "the quasi-religious aspect of the value of science" that was succumbing to the assaults of skepticism, and consequently, the contemporary scientist, unlike those in Russell's personal pantheon of scientific heroes, had submitted to the dogmas of the established order.[108] He cast the problem in terms of an opposition between science as the pursuit of power and science as the pursuit of truth; while the first becomes "increasingly triumphant," the second "is being killed by a scepticism which the skill of the men of science has generated."[109] It might seem that Russell was advocating a return to the era of scientific certainty which, if it could not be based on scientific facts, might be grounded on some metaphysical belief. Just as the argument seemed to be headed in this direction, however, Russell characteristically veered away, saying that he could not admit "that the substitution of superstition for scepticism advocated by many of our leading men of science would be an improvement."[110] Despite his association of science with religion, art, and love, however, Russell ultimately could not accept the solutions that people like Eddington proposed:

> Scepticism may be painful, and may be barren, but at least it is honest and an outcome of the quest for truth. Perhaps it is a temporary phase, but no real escape is possible by returning to the discarded beliefs of a stupider age.[111]

SUMMARY

The second theme to be found throughout Russell's publications in the interwar period, "the philosophical implications of the new physics," thus involved far more than discussions of mathematical relativity. Russell felt the need to challenge the metaphysical speculations of Eddington and Jeans in an unequivocal fashion, not because they were representative of physicists in general, but because their ideas were widely disseminated in the popular press. He was aware of what the scientific discoveries themselves implied, especially about what could be known about the physical world, but by 1931, Russell's primary concern seems to have been with how these discoveries were being used to buttress arguments in the popular press for the validity of religion in the universe revealed by the new physics. The public enthusiasm for the idea that the limitations of modern science left a place for religion thus led Russell to challenge various attempts by his contemporaries to explain how "reality" consisted of more than science could describe.

Russell's "conception of the nature of science," then, ties together the two themes of "the old savage and the new civilization" and "the philosophical implications of the new physics." Just as the development of modern science-as-technique posed the ethical question as to how the old savage might survive in the new civilization, the possibility of science being reconciled with religion posed an existential question, as to what the existence of the old savage might mean. Russell recoiled from both of these popular representations of science, asserting that the scientific outlook involved more than mere manipulation, on the one hand, and that it involved much less than metaphysics, on the other hand. Yet by the time Russell published *Religion and Science* (1935) and *Power: A New Social Analysis* (1938), it seems that he found a society created by scientific technique easier to accept than a cosmos explained by religion.

Eddington's assertion of a non-metrical universe, and the idea there was knowledge about existence other than that attainable through physical science, went to the heart of the debate over the relationship between science and religion. The terms of the debate had been set long before the interwar period, so that the new physics and other developments in modern science merely provided new evidence for Russell and his contemporaries to consider. In light of this new evidence, dogmatic conclusions from before the War about the primacy of either science or religion were rendered untenable. Russell's exchange with Eddington and Jeans was thus only a cameo in the larger drama played out in the popular press during the interwar period on the subject of science, religion, and reality.[112]

Chapter 5

Science, Religion, and Reality

One of the popular truisms of at least the last century has been "the conflict between science and religion." The idea of two opposing forces, or two protagonists doing battle, was as much an accepted viewpoint prior to the interwar period as was the Cartesian dichotomy between matter and mind. As the new physics dissolved the Cartesian dichotomy and as the Newtonian synthesis disintegrated, however, the inevitability of the conflict between science and religion came under question at the same time. Russell himself chose to accept "the conflict scenario" in a surprisingly uncritical fashion, given his skepticism on other intellectual fronts. Had he relied less on the prejudices of A. D. White, in particular, Russell might have found a way of reconciling epistemology and metaphysics that could have yielded some basis for the moral outlook he had wanted to promote. By the time of *Religion and Science* (1935), he had narrowed down his primary objections to the reconciliation between science and religion to the unfounded assertion that modern science provided evidence of cosmic purpose, and to the impossibility of acquiring knowledge about reality through mysticism or some other alternative to rational scientific inquiry. In the end, he found a cosmos explained by religion more difficult to accept than a society dominated by the utilitarian understanding of science-as-technique.

Before turning to *Religion and Science,* it is necessary to provide a theoretical context for discussing the relationship, conflictual or otherwise, between science and religion, and to illustrate the range of ideas in the popular press during the interwar period on that subject. Of all of Russell's book-length publications in this period, *Religion and Science* is the least able to stand on its own, without reference to the public conversation in which it was embedded. That public conversation on science and religion is characterized by a lack of precision in the definition of both terms, and a resulting lack of clarity in the range of conclusions reached by Anglo-American authors in the popular press. I contend, moreover, that the popular understanding of an inevitable historical conflict between science and religion is a recent development, a product of the reification of these terms, and of their abstraction from the historical and cultural contexts in which the accepted instances of that conflict took place.[1]

REIFYING SCIENCE AND RELIGION

To reify something is to regard something abstract as concrete, to attribute material status to an abstraction or real identity to a concept. The reification of Science and Religion, as indicated by the use of capital letters here, depicts them as separate and distinct entities, things that different people at different times would recognize as having more or less the same characteristics. By doing this, we can compare Science in nineteenth-century France to Science in sixteenth-century Egypt, or Religion in fifteenth-century England to Religion in Papua New Guinea. The process of reification abstracts the words "Science and Religion" from any single context and allows them to be compared between one culture or time period and another. We can discuss the history of the relation between Science and Religion from the time of Isaac Newton to the time of Stephen Hawking and assume that the words have more or less the same meaning regardless of their social, historical, or cultural context. This reification process thus implies a consensus as to what each term means, for without that consensus, any such comparison would be impossible.

The first difficulty with the reification of Science and Religion, however, is precisely that of the definition of terms. A cursory glance through dictionaries and recent histories of science will reveal that the language we use for "Science" is relatively recent in origin. When one looks at the different components of the social institution of contemporary Science—education, funding, laboratory settings, scientific publications, conferences, prizes, and so on—both the words we use and the activities they describe are products of (at most) the last 160 years. After Samuel Taylor Coleridge told the members of the British Association for the Advancement of Science in 1833 that they were no longer "natural philosophers" because of the exclusion of philosophy from their program, William Whewell, in response, coined the term *scientist,* which one scholar has observed was not actually in common usage until the twentieth century.[2] Thus it makes little sense to talk about anyone before 1840 as being a "scientist," for the word (and what it meant) did not yet exist. Similarly, what people did before 1840 should not be described as "science" either, for the social institution itself did not yet exist. Even a definition grounded in the popular perception of science as observation, hypothesis, experiment, and theory (the "scientific method") excludes far too many elements in the actual process of scientific discovery to be credible and may be more properly described as the root myth of modern science. Cross-cultural comparisons of science, or even comparisons within the same culture at significantly different times, are therefore fraught with historical inaccuracies that do little to clarify the relationship between Science and Religion.

When we consider a definition for "Religion," there also is much less clarity than the word itself might at first suggest. The "Religion" part of the "Science and Religion" discourse has little to do with any religion other than Christianity in Western Europe and North America and is as much a product of western culture since the sixteenth century as what we call Science. Furthermore, while there have been fewer changes in the nature of western Christian theology than in attitudes toward science in the last 500 years, we still cannot relate the theology of 1509 to the theology of 1859 and 1999 without considerable qualification. Add the differences in local religions due to social and cultural contexts, as well as the various explanations of religious experience yielded by the social sciences in the last 100 years, and a universal definition becomes impossible.

A reified Religion is thus more difficult to define than Science, even within the narrow confines of western Christian culture, and is still more difficult to define in a way with which different religious traditions would agree. From a secular perspective, there would be competing anthropological, sociological, and psychological definitions. To say that Religion deals with the metaphysical, defining it by exclusion from whatever is considered the "physical," is equally inadequate when actions and behavior are normally part of its expression. How Religion is defined requires choices to be made, and those choices in themselves contain a perspective on the subject of religious experience that is neither obvious nor necessary, more a creation of culture and history than anything else. For example, the social scientific and historical study of Religion is a product of late-nineteenth-century scholarship, with the increasing secularization of universities in Britain and the United States. Only when Religion began to be studied as a secular object in itself could there be the sort of equivalence between Christianity and other religious beliefs that would make its reification possible.

I argue, therefore, that a simple dualistic representation of Science and Religion only works in abstraction from the historical context in which actual science and actual religion—as institutions—developed in western culture since 1500. The period of the sixteenth to seventeenth centuries saw a self-conscious shift in some quarters from understanding knowledge of nature as one part of the spectrum of knowledge (including knowledge of medicine, knowledge of God, and knowledge of techniques), to what was described as "natural" or "mechanical" philosophy. Thus the divisions that would be recognized between "Science" and "Religion" today would have been incomprehensible at other times when much scientific activity was undertaken by people with theological training and interests. Studies of one prominent "natural philosopher"—Sir Isaac Newton—have rediscovered his interest in alchemy, for example, noting that he wrote more on theological topics than on the mathematical or "scientific" topics for which he is venerated. For Newton, as for others, the prospect of a dualistic understanding of the relation between Science and Religion would have been nonsensical.[3]

Yet these are the words we have been given to use, and much ink has been spilled in the last 140 years in defense of one, or in assailing the philosophical or moral defects of the other. We take for granted the existence of both Science and Religion. We assume these terms are sufficient for describing and explaining the human search for knowledge and meaning. We even tend to make the western Christian experience of reified Science and Religion into the universal standard against which the understanding of other cultures is measured and found wanting. Underneath the debate itself lies the further assumption that either Science or Religion, or some combination of the two, is capable of yielding Truth. The popular understanding of an inevitable historical conflict between Science and Religion is one product of the reification of these terms, and of their abstraction from the historical and cultural contexts in which the accepted instances of that conflict took place.

ORIGINS OF THE CONFLICT SCENARIO

It does not appear to be a coincidence that the origins of the "conflict scenario" between Science and Religion may be traced to the same period as their reification. There were two influential books in the nineteenth-century to which the conflict scenario may in large measure be attributed: J.W. Draper's *History of the Conflict between Religion and Science* (1874) and A. D. White's *A History of the Warfare of Science with Theology in Christendom* (1896). Both books went through multiple reprints, and while recent scholarship has demonstrated the factual errors and misinterpretations in Draper's and White's works, the conflict motif has continued to overshadow the popular perception of the historical relationship between religion and science to this day.[4] The hypostatization of this conflict, moreover, made it assume proportions beyond those that either author originally depicted.

Draper contended that the conflict resulted from the unbridgeable gulf between a divine revelation, which "must necessarily be intolerant of contradiction," and "the progressive intellectual development of man" due to "the irresistible advance of human knowledge."[5] The whole history of Science was for him "a narrative of the conflict of two contending powers." Draper's rhetoric came to characterize a common perception of the essential conflict between the discoveries of Science and the institutions of the Church:

> Institutions that organize impostures and spread delusions must show what right they have to exist. Faith must render an account of herself to Reason. Mysteries must give place to facts. Religion must relinquish that imperious, that domineering, position which she has so long maintained against Science.[6]

Draper, however, defined "religion" very narrowly; his argument was not with religion defined in social scientific terms, or even in terms of Christianity as a whole. For him it was not religion and science that were incompatible but "Roman Christianity and Science."[7] His diatribe was occasioned by the recent actions of Pope Pius IX and the Vatican Council in establishing papal infallibility. Draper's book was thus more parochial in scope than his sweeping judgments suggest; to substantiate his argument beyond this one concern, he advanced little historical evidence and a considerable amount of anti-Catholic rhetoric.

If J. W. Draper should receive the historical credit for being the first influential American author in the nineteenth century to advance the idea of the conflict between Science and Religion, Andrew Dickson White should be credited with providing the evidence to support such a thesis. He began his own polemic in 1869, when as the first president of Cornell University he delivered a lecture at Cooper Union Hall after coming under attack for refusing to impose a religious test on students and faculty. He recounted the famous "battles" between religion and science in the "persecution" of Copernicus, Giordano Bruno, Galileo, Kepler, and Andreas Vesalius, and he cited his own position as the latest victim of religion's war on science.[8] "Religion" became more narrowly defined by White as "ecclesiasticism" in his book *The Warfare of Science* (1876) and as "dogmatic theology" in the *History of the Warfare of Science and Theology in Christendom* (1896), the final form of his polemic.[9]

White said in his introduction to the 1896 volumes that he continued to write on the subject after Draper's book appeared because he became convinced the conflict was between "two epochs in the evolution of human thought—the theological and the scientific."[10] By this time, he considered the battle to have been won by Science. While it had been necessary to conquer "Dogmatic Theology," however, he believed Science would now go "hand in hand" with Religion.[11] According to White's analysis, the conflict thus seems to have been about power and control rather than about the nature of truth itself. As "theological control" continued to diminish, his own version of Religion would continue to grow stronger. To him, it was "'a Power in the universe, not ourselves, which makes for righteousness,' and . . . the love of God and our neighbor."[12] Here was an early example of what would emerge in a variety of forms during the interwar period: a definition of western "religion" somehow independent of the Christian tradition that had spawned it. White's opponents within the Christian church certainly would have resisted this kind of separation between the fundamentals of Christianity and their theology.

Although it is easy to find various examples of conflict in the history of Christianity and science, to make those conflicts a product of some metaphysical necessity as part of a master narrative is insupportable in terms of the historical detail. The conflict scenario, for example, ignored the fact that much of what passed for "science" had been done by people within the Christian church, who often made their living as clergy.[13] It also depended on a superficial understanding of the historic battles frequently referred to by its proponents from White onward.[14] To take the two best-known examples, neither the treatment of Galileo by the Catholic Church nor the reaction to Charles Darwin's theory of evolution by natural selection justify the simplistic interpretation placed upon them by proponents of the conflict scenario.[15] By the interwar period, moreover, Science and Religion had been reified in a way that tended to make their precise definitions incidental to the debate over their relationship to each other, and to the "reality" each ostensibly tried to represent.

SCIENCE AND RELIGION IN THE INTERWAR PERIOD

If what is meant by "religion" is a product of individual choice and not metaphysical necessity, such a plastic definition makes it possible for individuals with widely varying beliefs to find some common "religious" or "metaphysical" ground. There was a similar common ground for the popular perception of "science" after the Great War. E. E. Slosson and others popularized the "science digest," bringing ideas and theories into the public domain in a way that made them accessible to the intelligent lay reader.[16] Because of quantum mechanics, relativity theory, behavioral and Jungian psychology, sociology, and a host of other discoveries, science became something for general public consumption in the interwar period, even while its precise definition remained indistinct. Yet while the popular definitions of Science or Religion may have been vague in the interwar period, their relation was explicitly conceived in terms of the conflict scenario. Even if the grounds of the conflict were not clear, nor what each protagonist claimed in a given situation, nor how such an opposition could be applied to a specific subject (such as the new physics), the perceived opposition between religion and science continued to dictate the terms of the popular debate for all participants, including Bertrand Russell. This dualism dominated the interwar period discussions on science and religion, regardless of whether an author was promoting reconciliation, coexistence, or an irresolvable conflict.

In the perception of Draper and White, the conflict between science and religion would ultimately always find science victorious, as various superstitions and dogmas maintained by the Church were replaced by theories based on hard scientific evidence. Religion was represented, at least in its institutional form, as an impediment to the acquisition of knowledge about nature and a barrier to scientific progress itself. In the language of materialism, religion dealt with the subjectivity of emotional experience, not the objectivity of the physical world. Where one could ascertain truth about the physical world by proper scientific method, no such truth could be forthcoming from a subjective emotional experience.

With the advent of relativity theory and the resultant demise of absolute materialism, the arguments used against the possibility of acquiring knowledge or truth through religion could as easily be used against science. The neat Cartesian dichotomy between mind and matter was replaced by a kind of interpenetration between the two which, as it gave credence to the discoveries of science, threatened to give similar credence to the assertions of religious belief. One of the first results of this shift in understanding due to the philosophical implications of the new physics was to rehabilitate "religion." The dualism inherent in the conflict scenario led, in the interwar period, to a general acceptance of the idea that the realm of science was juxtaposed with the realm of religion, however each was constituted. This led to numerous attempts to characterize the realm of religion as not only distinct from the realm of science but also as somehow equivalent. This equivalence in turn led to a surprisingly strong affirmation of the value of religion, both to the individual and to society—noting, of course, that while religious belief was affirmed, institutional religion often was not. This led to the idea that whatever boundary might be established between religion and science, there was some mutual influence that was beneficial to both parties, and to the society wise enough to promote reconciliation on such terms.

These alterations to the pre-War understanding of the conflict scenario led to three general approaches to the relationship between science and religion: first, the two realms were distinct, and a peaceful coexistence was to be established with neither side meddling in the affairs of the other; second, there was a reconciliation to be effected through transcending the limitations of both science and religion, as modern science had revealed that both describe complementary aspects of the same reality; and third, as science purges the outdated dogmas and superstitions from religion in the modern world, religion will provide what is missing in a society ruled by scientific technique.

To illustrate these points from the literature of the interwar period, en route to a discussion of Russell's *Religion and Science,* we now focus on two influential volumes of essays, with supporting examples cited from among the numerous books and articles on the relationship between science and religion that reflect an intense public interest. *Science, Religion and Reality* (1925)[17] was intended to be an explication of the subject, not an apologetic for Christianity or an attack upon it. Its contributors were prominent in their fields; among them were A. J. Balfour, who wrote the introduction,[18] Dean W. R. Inge, who wrote the conclusion,[19] Bronislaw Malinowski, whose famous "Magic, Science and Religion" explored the subject from an anthropological perspective, and Arthur Eddington, whose "The Domain of Physical Science" gave a physicist's view and set the stage for his subsequent Gifford Lectures.[20] *Science, Religion and Reality* did not pander to any one resolution of the conflict between science and religion, and Russell (who at least read Eddington's contribution, if not others[21]) was not moved to counter in print any of the essays it contained. A second influential collection of essays reached a different conclusion and received a different response from Russell. Between September and December 1930, a series of BBC lectures resulted in another anthology, published as *Science and Religion.* The purpose of the series was to give

> a personal interpretation of the relation of science to religion by speakers eminent as churchmen, as scientists, and as philosophers; and to determine, in the light of their varied and extensive knowledge,

to what degree the conclusions of modern science affect religious dogma and the fundamental tenets of Christian belief.[22]

The question they dealt with was how the reality presented by science affected Christian theology. Unlike the earlier book, the contributors had the opportunity to hear the broadcast of each other's ideas, and they had been able to revise their chapters before publication. The result was a more coherent collection of shorter, less technical papers on a narrower theme, one that at the same time allowed more explicitly for the presentation of personal opinions. These papers tended to resolve the conflict between Science and Religion in a congenial manner, maintaining the dualism inherent in the conflict scenario, but allowing Religion importance in its own right.[23] Russell was irritated by such congeniality, thus providing the impetus for his response in *Religion and Science,* the latter chapters of which dealt specifically with *Science and Religion.*

In terms of the "conflict" between science and religion, if religion and science dealt with separate realms of knowledge, truth, and experience, then there could be conflict only when either departed from its territory. For example, biologist J. Arthur Thomson wished to demonstrate "that an opposition between scientific description and religious interpretation is fundamentally a false antithesis."[24] He stated that "the aims and moods are quite different, and there is no justification for what has been called 'warfare' or 'conflict.'"[25] If there were any boundary disputes, he said they could be settled through the arbitration of "a frontier commission," because we had to learn "to render unto Science the tribute that is its due; and to God the things that are HIS."[26] Theologians such as E. W. Barnes, the Bishop of Birmingham, on the other hand, wished science to "renounce all pretensions to metaphysical dogmatism,"[27] leaving that field to a religion that finds God's nature revealed in the Universe as a whole.[28]

The majority of contributors to *Science, Religion and Reality* similarly favored the dualism of separate realms for religion and science. Referring to Draper's *History of the Conflict between Religion and Science,* Balfour observed in his introduction that by 1925, Draper's notion of the apocalyptic end of religion at the hands of science was "wholly without warrant" in western culture.[29] Nor was Balfour happy with Draper's version of the "conflict scenario." He said that the idea of an impending battle to the end between science and religion "seems to be neither good philosophy, nor good theology, nor good science,"[30] and he opted instead for "an unresolved dualism." Balfour claimed that "our experience has a double outlook," both material and spiritual, continually influencing each other, though he could not say what the boundary between them might be.[31]

Dean Inge wrote in his concluding essay that the object of *Science, Religion and Reality* was "to make clear what the present state of the relations between science and religion actually [was]," setting out to accomplish the "practical object" of "indicating possible terms of peace, or a *modus vivendi*."[32] As such, the book presented a range of possibilities and not a resolution. Nor did Inge attempt such a resolution, although he specifically dismissed the idea of a division of territory as a means to end the dispute:

> A religion which does not touch science, and a science which does not touch religion, are mutilated and barren. Not that religion can ever be a science, or science a religion; but we may hope for a time when the science of a religious man will be scientific, and the religion of a scientific man religious.[33]

Having read all of the essays in this book, Inge was "inclined to feel confident that a reconciliation is much nearer than it was fifty years ago,"[34] but he himself would not advance such a conclusion.

This was by no means a generally accepted conclusion, even among the ranks of the clergy. Harry Emerson Fosdick, for example, derided the idea that "real" science and religion would not conflict:

> Some vaguely progressive minds take too much comfort in such consoling generalities as that true science and true religion cannot conflict. The proposition is so harmless that no one is tempted to gainsay it but, so far from solving any problems, it serves only to becloud the issue. The plain fact is that, however true science and true religion ought to behave toward each other, actual science and actual religion are having another disagreeable monkey-and-parrot time.[35]

Walter Lippmann, in his *A Preface to Morals* (1929), also was unsympathetic to the idea of such a simplistic reconciliation: "While the policy of toleration [between religion and science] may be temporarily workable, it is inherently unstable."[36] That instability was a result of the abdication of the essence of religion, a denial of the validity of revelation in favor of a methodology that could not yield the truth that was supposed to be the essential character of religious experience:

> A reconciliation . . . may soften the conflict for a while. But it cannot for long disguise the fact that it is based on a denial of the premises of faith. If the method of science has the last word, then revelation is reduced from a means of arriving at absolute certainty to a flash of insight which can be trusted if and when it is verified by science. Under such terms of peace, the religious experiences of mankind become merely one of the instruments of knowledge, like the microscope and the binomial theorem, usable now and then, but subject to correction, and provisional. They no longer yield complete, ultimate, invincible truths. They yield an hypothesis.[37]

Some other means of resolving that boundary dispute had to be found if religion was to be of value to humanity in the modern world. Simple toleration or a reconciliation based on inappropriate assumptions would not suffice to do this, because the essential value of religious experience would thereby be lost.

Two other contributors to *Science, Religion and Reality*, Antonio Aliotta and John Oman, made some attempt to deal with the boundary question by reconciling science and religion in terms of a higher level of truth, one that transcended the duality inherent in the conflict scenario. Aliotta urged his readers to put aside "the old intellectualistic conception of reality as a thing in itself" and the correspondence theory of truth between our ideas and that reality, which he called "a desperate undertaking."[38] Instead, he asserted the holistic nature of a religious outlook on life, one in which reality is never exhausted by the ideas which we have of it, because "the concrete unity of experience" is dynamic rather than static.[39] Oman, in his article "The Sphere of Religion," began by asserting that "because the world is one and known to our minds as one universe of discourse, no subject of study has absolutely determined frontiers."[40] Notwithstanding this opinion, Oman went on to define the sphere of religion as dealing with what Rudolf Otto earlier had called the "numinous," or the "holy."[41] In response to the charge that religion dealt not with facts but with metaphysical speculation, he asserted that it dealt with "feeling and value" in no more subjective a way than with the ostensibly material aspects of "the visible world."[42]

In his conclusion to *Science, Religion and Reality,* Inge said the Church had three options with respect to the discoveries of modern science: first, to declare them "impious and heretical" (an attitude that fueled public perception of the antiquated nature of institutional religion); second, to claim a higher truth than science could attain, which when pushed to its extreme takes the form of "acosmism" ("the theory which denies the objective existence of the world or universe"); or third, to "recast" all theological doctrines dependent on theories which science has disproven.[43] He advocated the third option, as a result of which the Church would be forced "to think of God less anthropomorphically, and of heaven as a state rather than a place."[44] Inge was confident that this was the right course to take, for he did not think that "the religion of Christ" was dependent on any "false opinions of the nature of the universe." He urged the Church to face this crisis "candidly," with "faith and courage" and "common honesty," and he asserted that it could prevail if it did so: "Only let us hear no more of clergymen thanking God that theology and science are now reconciled, for unhappily it is not true."[45]

Similar conclusions were reached by very different individuals during the interwar period. J. B. S. Haldane had a less sympathetic view of religion, for he believed it was "full of obsolete science of various kinds, especially obsolete cosmology and obsolete psychology,"[46] and he blamed this problem for its decline. He was "rather inclined" to think there was some core in religion that was "independent of scientific criticism," though he did not say what he thought that might be.[47] Alfred North Whitehead noted with approval that science helped eradicate such obsolete beliefs from religion:

> Religion will not regain its old power until it can face change in the same spirit as does science. Its principles may be eternal, but the expression of those principles requires continual development. This evolution of religion is in the main a disengagement of its own proper ideas from the adventitious notions which have crept into it by reason of the expression of its own ideas in terms of the imaginative picture of the world entertained in previous ages.[48]

In other words, religion and science might coexist, but only if the adherents of religion accepted the need for religious beliefs to evolve in a way that was consonant with the current scientific worldview. The essays in Shailer Matthews's book, *Contributions of Science to Religion,* indicated a similar belief.[49] In his final section on "Religion, the Personal Adjustment to Environment," he wrote on "The Evolution of Religion," "Scientific Method and Religion," "Science Justifies the Religious Life," and "Science Gives Content to Religious Thought."[50] He demonstrated that, because religion is part of human culture, it has evolved as that culture has evolved. Just as there was no final state of human culture, there also could be no final state of religion, he maintained. Science, therefore, provided the means by which religion was enabled to evolve through the elimination of beliefs and ideas no longer appropriate to the changed environment in which humanity found itself.[51]

PERSONALISM AND A DUAL REALITY

Whatever the talk of reconciliation between science and religion, the influence of the conflict scenario helped perpetuate their duality. This duality also was in part the result of the philosophy of "personalism," which was in vogue during the interwar period. While there were various forms of personalism, at its core was the belief that reality could best be interpreted in personal terms, making the idea of "personality" its central precept.[52] Personalism could take theistic as well as atheistic and pantheistic forms, but its metaphysics "may be summed up in the statement that personality is the key to reality."[53] In saying this, it maintains the "autonomous validity" of religion as a distinct realm of human understanding, and "it distinguishes so clearly and sharply between science and religion that logically there is and can be no real conflict between them."[54] Personalism distinguishes between phenomenal and ontological realities, thereby eliminating the possibility of conflict when the significance of this distinction is accepted.[55]

The prevalence of this philosophy in the debate over the relationship between science and religion in the interwar period is best demonstrated in the work of several of the contributors to the second significant anthology in the interwar period, *Science and Religion.*

In Arthur Eddington's view, the difference between science and religion was not between "the concrete and the transcendental but between the metrical and the nonmetrical."[56] There was thus a spiritual as well as a physical world, and he outspokenly repudiated the idea "of proving the distinctive beliefs of religion from the data of physical science" or by its methods.[57] Religion was for him something "mystical," based not on science but on "a self-known experience."[58] Thus, he said, "We have to build the spiritual world out of symbols taken from our own personality, as we build the scientific world out of the symbols of the mathematician."[59] Eddington acknowledged the criticism of those who wanted him to defend the idea of an objective reality of God but claimed that we only know God in the same way we know about the physical world, which is through a kind of experience that is, by nature, subjective.[60]

For J. S. Haldane, the standpoint of religion was the realization of the significance of personality, which served, in the end, to make science part of religion:

> When we realise that it is personality which unites us all, and unites and includes the whole of our experience, we call that personality God, and join in looking up to God and seeking to do His will as revealed to us in our experience. Science, for instance, as pursuit of truth, even if it be only relative truth, is thus part of religion.[61]

Haldane, however, was not exactly orthodox in his religious beliefs; he repudiated the "professed beliefs of religious organizations" because they were bound up with materialistic ideas of causality, and he said "beyond and around them all there is a Church of God, which all may belong to, and which is associated with neither materialism nor belief in miracles."[62] Consequently, he maintained, both religion and "real Christianity" are based on "experience," in a universe that is "no mere physical or biological universe, but a spiritual universe of values which can only be expressed as the active manifestation or personality in an indefinite or chaotic background."[63] This resulted in the notion of "the unifying and all-embracing Personality of God."[64]

Haldane found support for a dualistic view of the universe in the distinction between mechanistic and biological interpretations,[65] saying that "biology bars decisively the door against a final mechanistic or mathematical interpretation of our experience" in a way that allows for ideas about religion.[66] Because notions of truth are not merely individual, he points to the obvious existence of some shared ideal of what "truth" means, and the value that is to be placed on it.[67] Therefore, he concludes, "personality is not merely individual," making God not merely "a being outside us, but within and around us as Personality of personalities."[68] Repudiating revelation and natural science alike as sources of religious belief, Haldane maintained: "It is only within ourselves, in our active ideals of truth, right, charity, and beauty, and consequent fellowship with others, that we find the revelation of God."[69] The Personality of God becomes the unifying force in the universe, for "the only ultimate reality is the spiritual or personal reality which we denote by the existence of God."[70]

Julian Huxley shared J. S. Haldane's perception of the unifying force of "personality" in the universe, though he had some important qualifications about the nature of religious experience. He said, "Religion itself is the reaction between man as a personality on the one side, and, on the other, all of the universe with which he comes in contact."[71] By organizing "our knowledge of outer reality after the pattern of a personality," we make possible its interpenetration of "our private personality."[72] "To my mind," Huxley stated, "[this] is what actually happens when men speak of communion with God."[73] The operation of religion, then, becomes "a relation of the personality as a unit to external reality as a unit—and a relation of harmony."[74]

Huxley stated he was an adherent of no religion but believed "religious feeling is one of the most powerful and important of human attitudes."[75] He mentioned the significance of the "science of religion" in identifying common elements (a sense of sacredness, a sense of dependence, and desire for explanation and comprehension), while affirming the applicability to religion of the "principle of development." Like A. D. White, Huxley perceived the conflict to be between science and theology, not science and religion; theology became incorrectly "tinged" with the sacred and made science seem in conflict with particular stages of particular religions. Huxley said the man of science "worth his salt" has a "definitely religious feeling about truth," and that religion must assimilate scientific discoveries. He described science as "partial, morally and emotionally neutral," and he depicted the practical task of religion to be finding the best use for (scientific) knowledge and powers.⁷⁶ Throughout his essay in *Science and Religion,* Huxley recounted the same ideas that were to be found in some of his other books, such as *Religion without Revelation,*⁷⁷ considering the essence of religion to be some reified quantity abstracted from traditional western religious experience, which could best be expressed as an aspect of both individual and cosmic "personality."

The hierarchy implicit in the personalist attitude to science and religion places the truths revealed through religion above those discovered by science. Russell was not alone in objecting to such a hierarchy, nor in objecting to the idea of a sphere for religion that was distinct or separate from that of science. Yet the problem of whether it was possible to distinguish between a phenomenal and an ontological reality persisted. Whereas Russell would say it was impossible to *know* whether an ontological reality existed or not, others would say it was possible to have a mystical *experience* of such a reality. Therefore, the question of whether individual experience constituted some form of knowledge, and whether religion was therefore a means to discovering truth, lay at the center of the supposed reconciliation between science and religion.

In summary, there seemed to be a consensus in the popular press during the interwar period that the spiritual needs of the scientific society could not be met by antiquated theology, nor was science by itself able to address the questions of meaning and purpose that its discoveries, and the experience of life in the Machine Age, provoked. Among Russell's contemporaries, Edwin Burtt's conclusion was one of the more succinct:

> Religion will find itself in the modern world when it envisions its object of worship in terms that square with the human values for which science stands. A religion and science thus brought together in one is the deepest need of the modern world.⁷⁸

It is fair to say that there was both an existential and an ethical impulse to effect such a reconciliation: meaning expressed either in only religious or scientific terms was held to be inadequate, while a society constructed by science, without the meliorating influence of religion, was unlikely to be one in which people would want to live.

RUSSELL ON RELIGION AND SCIENCE

Russell had a very low tolerance for discussions about the congenial relationship between science and religion. When J. S. Haldane pronounced that God was the ideal "Personality of personalities," for example, Russell retorted: "Statements of this kind, I must confess, leave me gasping, and I hardly know where to begin."[79] First of all, he had no use for the idea of equivalence between the realms of science and religion; for Russell, there was only one world, and that was the one revealed through the methods of science and manipulated through the operations of scientific technique. Whatever the relation between science and religion, it could not be one in which some other sphere or realm was postulated for religion. Second, Russell was vehement in saying that religion had nothing to offer science, and that the beneficial effect science had on religion was to eliminate the false beliefs and superstitions by which religion was constituted. As for the relation between science and religion, his own work reflects little sympathy for anything other than the conflict scenario. The idea of the peaceful coexistence between two realms foundered on the practical problems of demarcation; not only was there only one world, he believed, but science itself had dissolved the traditional boundary between mind and matter. It was somewhat absurd to assert any other essential dichotomy when science seemed to be on the brink of understanding the interpenetration of mind and matter, to the point that each was an expression of the other, or both of some third "neutral stuff." Regarding the reconciliation of science and religion in terms of some higher level of truth, Russell was incapable of seeing how the provisional truth claims of science might ever be reconciled with the dogmatic assertions of Christianity, let alone those of other religious traditions. Even if an individual could be both a scientist and a religious person, the institutional expression of religion and the uniformity of thought required through the acceptance of creeds rendered any true reconciliation impossible.

From the opening chapter of *Religion and Science* (1935), Russell made clear the context in which he viewed the relation between the two. Entitled "Grounds of Conflict," the chapter begins with his categorical statement that, "between religion and science there has been a prolonged conflict in which, until the last few years, science has invariably proved victorious."[80] Though he went on to identify German fascism and Soviet communism as the two "religions" that threatened the victorious record of science in this conflict, the stated purpose of his book was "to examine the grounds and the history of the warfare waged by traditional religion against scientific knowledge."[81] His concern was not with science or religion in general but with "those points where they have come into conflict in the past, or still do so at the present time."[82] The influence of Andrew Dickson White is obvious in comments such as these; in a book with few footnotes, Russell cites *The Warfare of Science and Theology* three times.[83]

As much as Russell himself tended to reify ideas such as Freedom, Organization, Power, or Religion and Science, he also made an effort to analyze the idea that he had abstracted from its historical and cultural context. Thus, his understanding of Religion ("considered socially") was separated into three aspects: the Church, creeds, and personal moral code.[84] In other words, he considered the Church as a social institution; creeds as dogmatic expressions of collective belief; and the individual's expression of religion in terms of personal beliefs and actions. This focused his analysis of the relationship between Religion and Science on three areas: institutional competition based on the social structures of power; epistemology and competing claims to truth; and the personal application of morals to ethics, or of values to individual actions.[85] Russell saw the conflict between Religion and Science as one that centered primarily on the second area, epistemology and competing claims to truth. Taking his cue from A. D. White, Russell considered creeds "the intellectual source of the conflict between religion and science," and "the bitterness of the opposition" to be due to their connection to the Church and to moral codes:

> Those who questioned creeds weakened the authority ... of Churchmen; moreover, they were thought to be undermining morality, since moral duties were deduced ... from creeds. Secular rulers, therefore, as well as Churchmen, felt they had good reason to fear the revolutionary teachings of the men of science.[86]

This was especially problematic when the conflict involved something deeper than a dispute arising specifically out of the Bible, such as when science "controverts some important Christian dogma, or some philosophical doctrine which theologians believe essential to orthodoxy."[87]

While the Christian perspective on the truth-values of its creeds had an intellectual and a philosophical foundation measured in centuries, a coherent depiction of the perspective of modern science was much more elusive. For Russell, the key point was how the knowledge expressed in theological and scientific truths was obtained. For him, a claim to absolute truth required absolute knowledge. Russell rejected the idea that knowledge might be gained through mystical experience, because "knowledge" had to be verifiable. Science, on the other hand, made provisional truth claims about the nature of reality and allowed for the adjustment of those claims as the acquisition of new knowledge might require. Russell saw conflict occurring in those historical instances where the truth claims of Religion were incompatible with the truth claims of Science. Thus he was not advocating the absolute truth understood through Science over the absolute truth revealed through Religion. Rather, he was protesting the illegitimacy of the conclusions about absolute Truth reached by scientists and theologians alike about what could be known about the nature of the physical world, and about the meaning of life itself.

This was a concern Russell had expressed before the Great War. He was resolutely opposed to any attempt to prove the existence of God from science, as illustrated in his review of A. J. Balfour's first set of Gifford Lectures, published in 1915 as *Theism and Humanism*.[88] Not waiting for the published version, Russell assailed the newspaper accounts of Balfour's lectures in a paper published in *The Cambridge Review*. Balfour adduced the existence of God from the existence of a priori beliefs in science that "are not self-evident, not logically inevitable, not derivable from experience, and yet such as no one is willing to abandon."[89] By criticizing science in this way, he intended to demonstrate that it rests on beliefs which, "unless theism is assumed, must appear baseless and unwarrantable."[90] In letters to Ottoline Morrell, Russell's reaction to Balfour's lectures was more pungent than his review: Calling the lectures "rhetorical sentimental twaddle"[91] and "incredible balderdash," he said, "There is something about every word of his writing that fills me with loathing. I suppose he is not utterly vile, but I feel as if he were."[92]

Similar feelings also characterized portions of *Religion and*

Science, which was written as his objection to the BBC series published in 1930 as *Science and Religion*. Russell's objections to the reconciling spirit of the essays in *Science and Religion* were based upon a close reading of the text. The notes he made while he read it are preserved in the Russell Archives and consist mainly of verbatim quotations; most of these same quotations appeared in the two chapters of *Religion and Science* that he devoted to rebutting the arguments of their authors. As a result, these notes and Russell's own book may be used as his gloss on the BBC series.[93] His sardonic remark that "outspoken opponents of religion were, of course, not included" might be seen as the starting point for his public rejoinder. Where Canon B. H. Streeter commented on the recurrent idea in the series "that science by itself is not enough," Russell questioned "whether this unanimity is a fact about science and religion, or about the authorities who control the BBC."[94]

The first six chapters of *Religion and Science* set out Russell's version of the historical conflict between the two protagonists, taking as his examples the advent of the Copernican theory of heliocentrism, the "persecution" of Galileo, the controversy over evolution, and the struggle between scientific medicine and "demonology." He then dealt with the lack of proof for the existence of the soul, and for what modern psychology considered "personality." His sixth chapter was a reprise of his earlier published arguments on determinism, causality and free will in light of quantum theory. It was not until chapter 7 of *Religion and Science* ("Mysticism") that Russell focused on the issue of the supposed reconciliation between science and religion (depicted by every contributor to *Science and Religion* except Julian Huxley) and how this might be related to a modern view of mysticism.

MYSTICISM

In *Religion and Science,* Russell continued to reject a reconciliation between science and religion, especially if it came out of the abdication of science's right to explore the boundaries of reality. He focused his objections on the opinions of three scientists regarding the limits of science: J. Arthur Thomson noted that "science as science never asks the question Why."[95] J. S. Haldane observed that it is "only within ourselves . . . that we find the revelation of God."[96] Bronislaw Malinowski declaimed that, "religious revelation is an experience which, as matter of principle, lies beyond the domain of science."[97] While Russell agreed with the idea that "science is not enough," he did not accept its extension to postulating the existence of some other nonscientific means of arriving at truth. Quotations such as these indicated a general sense that the limitations of science required some other means than the scientific method to discover the truth about what Eddington called "the non-metrical universe." Russell, however, believed these three scientists were making the same mistake as Eddington and Jeans in trying to arrive at conclusions about "reality" that were based on metaphysical speculations.

In his own discussion of mysticism, Dean Inge was more explicit about the extra-scientific means of acquiring knowledge of the "non-metrical world," saying that "the proof of religion . . . is experimental."[98] Whereas science "deals with facts," he said, religion "deals with values." Although both are "real," they are on "different planes," and both cannot help "poaching" on the other's preserve. Russell interpreted this statement of Inge's to mean that "religion must make assertions about what is, and not only about what ought to be."[99] He then gave serious consideration to Inge's suggestion that the unanimity of mystics of different religions gave credence to the "proof" of religion's claims about truth, referring to his earlier discussion of it in *Mysticism and Logic*.[100] On the matter of truth being ascertainable through revelation or through the revelatory experiences of mystics, however, Russell disagreed. He believed the assertions of the mystic about the nature of the universe were "inessential," saying, "I cannot admit any method of arriving at truth except that of science, but in the realm of the emotions I do not deny the value of the experiences which have given rise to religion."[101] He concluded his chapter on mysticism with words reminiscent of Andrew Dickson White, commenting that the association of religious experiences with "false beliefs" has led "to much evil as well as good." When they are freed from this association, by properly ascribing truth about the nature of the universe only to science, "it may be hoped that the good alone will remain."[102]

Russell's separation of knowledge from experience, at least in the realm of religion, was his attempt to stay out of the instrumentalists' camp when it came to understanding the relation between science and religion. By allowing religious experience to

Science, Religion, and Reality

have an emotional validity, he tried to leave room for a personal metaphysic without having to accept the collective truth claims advanced in the form of dogma by the Church. As Russell had recounted in the opening chapter of *Religion and Science,* the perception of religion as personal feeling was the only aspect of his threefold division of "religion" (into institution, creeds, and a code of personal morals), which he was prepared to accept as "independent of the discoveries of science":

> The man who feels deeply the problems of human destiny, the desire to diminish the sufferings of mankind, and the hope that the future will realize the best possibilities of our species, is nowadays often said to have a religious outlook, however little he may accept of traditional Christianity. In so far as religion consists in a way of feeling, rather than a set of beliefs, science cannot touch it.[103]

The dilemma posed by "the old savage in the new civilization," however, would not allow a purely personal religion to be of either value or use in a society searching for a collective ethic to ensure its survival. Russell's concession of the emotional validity of personal religious experience, in the end, was thus of little consequence in the resolution of this dilemma.

COSMIC PURPOSE

The eighth chapter of *Religion and Science* ("Cosmic Purpose") went on to tackle what Russell obviously considered a root problem, not only of the BBC series but of contemporary discussions about the problem of meaning with respect to the recent discoveries of science. Both modern men of science ("if they are not hostile or indifferent to religion") and liberal theologians alike "cling to the one belief which, they think, can survive amid the wreck of former dogmas—the belief . . . in cosmic purpose."[104] His notes on *Science and Religion* began with a schematic that he then followed in this chapter, laying out the three forms of the dogma of cosmic purpose, all of which "have in common the conception of Evolution as having a direction towards something ethically valuable."[105] They were: the theistic, that God created the world and the laws of nature toward some purpose in the mind of the Creator, "who remains external to His creation"; the pantheistic, in which God is "merely the universe considered as a whole" and thus the act of creation is replaced by a creative force within the universe that causes it to develop according to its plan; and the "emergent," which Russell said was "more blind" in its purpose, with an end explicit in the beginning only in "some rather obscure sense."[106] Bishop Barnes was set out as the advocate of theism, J. S. Haldane of pantheism, and Samuel Alexander of the "emergent" school, although Russell also acknowledged (and discussed) its "more typical representatives," Henri Bergson and Lloyd Morgan.[107]

Though Russell explicated each of the three options for explaining cosmic purpose at length, and in the authors' own words, his dismissal of them was cutting and abrupt. The theistic standpoint foundered on its inability to account for the existence of evil in a universe created by an omnipotent God;[108] to Haldane's assertion there was more to life than physical and mechanical processes, Russell cited the general lack of support for his position and belittled the notion of "personality" as sufficient for any kind of explanation of matter or of God, thereby also dispatching the philosophy of personalism.[109] As for the "emergent" school, Bergson was dismissed as a "poet" whose evidence consists of "occasional bits of bad biology,"[110] and to Russell the idea as a whole was based on an argument that was "extraordinarily thin."[111] Russell then tackled the whole question of cosmic purpose itself, grounding his criticisms in the two statements he used to characterize anyone who held this belief: first, that "those who believe in Cosmic Purpose always think that the world will go on evolving in the same direction as hitherto"; and second, that "what has already happened is evidence of the good intentions of the universe." Both of these propositions he felt were "open to question."[112] Not only was there no reason to suppose "evolution" in the universe would continue as it had in this small corner of it (referring to James Jeans's work on cosmology), Russell also lampooned those who believed in the benign character of the universe simply because it produced humanity:

> Man, as a curious accident in a backwater, is intelligible: his mixture of virtues and vices is such as might be expected to result from a fortuitous origin. But only abysmal self-complacency can see in Man a reason which Omniscience could consider adequate as a motive for the Creator.[113]

SCIENCE AND ETHICS

Russell's repudiation of mysticism as anything more than individual emotional experience and his dismissal of the various attempts to derive a sense of cosmic purpose from the operations of science set up the final chapter of *Religion and Science*. The kind of language found in the concluding chapter of *The Scientific Outlook* ("Science and Values") had disappeared by 1935. In *Religion and Science,* Russell asserted an absolute disjunction between facts and values, assigning facts (and therefore knowledge and truth) to the realm of science and values to the realm of ethics and experience.[114] He felt that science had nothing to say about values, and he objected to the idea that ethics contained truth "which cannot be proved or disproved by science."[115] The stages in Russell's argument about what followed from this assertion merit close attention.

He began by dividing ethics into moral rules and "what is good on its own account."[116] Moral rules are adjudged good or bad by the consequences when they are applied. Questions regarding"values," however ("what is good or bad on its own account, independent of its effects"), were "outside the domain of science."[117] Russell drew the further conclusion that values were outside of the domain of knowledge altogether and were an expression instead of emotion.[118] He considered ethics an attempt to escape from the subjectivity that this definition of values entailed, and thus ethics was "closely related to politics," "an attempt to bring the collective desires of a group to bear upon individuals" or the attempt by an individual "to cause his desires to become those of his group."[119] Therefore, Russell said, ethics was "an attempt to give universal, and not merely personal, importance to certain of our desires."[120] This may be attempted from either the point of view of the preacher, or of the legislator, he said. The legislator compels the adoption of a given desire by the group, while the preacher has to generate an equivalent desire in the minds of his audience.[121] Russell observed, moreover, that the confusion that existed in ethics was the result of the "interlocking" of the personal wish, with the desire for something universal.[122] Thus, he said, ethics ultimately could not contain statements about what was true or false but only about desires. Science might discuss the causes of desires, or how they might be realized, "but it cannot contain any genuinely ethical sentences, because it is concerned with what is true or false."[123] He called for the cultivation of "large and generous desires" to achieve moral improvement, repudiating any objective or intrinsic values.[124]

In this chapter, Russell sealed the fate of his "moral outlook in science" and the failure of his attempt to solve, in something other than utilitarian terms, the dilemma posed by the survival of "the old savage in the new civilization." If ethics was only an expression of desire, not of any kind of knowledge or truth, then there was no reason to assume that any group of individuals, much less an entire society, would share equivalent desires, thereby rendering a social ethic by consensus impossible. The attempt to give universal expression to personal desire could be undertaken by the legislator or the preacher, but while the legislator had the machinery of government (including propaganda) to produce conformity, the preacher had only emotion and rhetoric. If values have no intrinsic character, if there is no metaphysical verity behind their expression, then they are rendered impotent by the political and economic conditions that shape the desires of different groups in society. Russell might call for the cultivation of "large and generous desires," but there was no good reason for anyone to listen.

Russell's acceptance of the conflict scenario thus meant for him that religious experience could yield neither knowledge or truth. Religion, or religious experience, might be valuable at a personal level, but it lacked the universal character it needed to be useful to society (in terms of a foundation for ethical principles). Without that universal character, it also could not yield an understanding of reality that was distinct from that already provided through the operations of science. For Russell, the new physics yielded "negative knowledge" about what could not be known through science, and about the provisionality of what knowledge could be acquired, but it also eroded any grounds for making mysticism more than a purely personal experience. As he recorded in his *Autobiography*, there were emotional consequences to living in a world bereft of universals because of the new physics, consequences that were a result of his retreat from Pythagoras after the Great War:

> I had been very full of personal misery in the early years of the century, but at that time I had a more or less Platonic philosophy which

enabled me to see beauty in the extra-human universe. Mathematics and the stars consoled me when the human world seemed empty of comfort. But changes in my philosophy have robbed me of such consolations. Solipsism oppressed me, particularly after studying such interpretations of physics such as that of Eddington. It seemed that what we had thought of as laws were only linguistic conventions, and that physics was not really concerned with an external world. I do not mean that I quite believed this, but that it became a haunting nightmare, increasingly invading my imagination.[125]

In a world where truth, and knowledge of that truth, was provisional and not absolute, meaning and purpose could only be expressed in terms of personal desire or belief. Universals of any sort could have no existence of their own, apart from the mind of the individual who impressed them on an indifferent cosmos. A creed or an ethic might be promoted in society, but its general acceptance depended upon rhetorical persuasion or social coercion, for it could express neither knowledge or truth. As a consequence, the moral outlook in science, which he had tried to articulate in opposition to the instrumentalist understanding of science-as-technique alone, became untenable. Russell was left with no practical response to the dilemma posed by the survival of the old savage in the new civilization other to focus on finding some way to understand and to control the political and economic effects of power in the modern world. It is a "mad world" after 1914, he concludes in *Education and the Social Order* (1932), one that "has become so intolerably tense, so charged with hatred, so filled with misfortune and pain that men have lost the power of balanced judgment which is needed for emergence from the slough in which mankind is staggering."[126] Trying to convince himself as much as his audience, he wrote:

> Our age is so painful that many of the best men have been seized with despair. But there is no rational ground for despair: the means of happiness for the human race exist, and it is only necessary that the human race should choose to use them.[127]

As much as anyone else, Russell wanted these words to be true, but his rhetoric was not powerful enough to persuade someone who knew only too well the inherent irrationality of the old savage in the new world being created by the applications of science.

Chapter 6

Science and Power

By the time he published *Power: A New Social Analysis* in 1938, Bertrand Russell was reaping the whirlwind he had sown since the end of the Great War. He was well aware of the dangers of organization as applied to the State in the Machine Age and of how a new outlook was necessary to avoid the Next War that seemed the inevitable result of economic nationalism. He understood what the "scientific society" would mean for humanity, should the individual be totally subordinated to the group. He also could see the consequences of the desire for power when it was exercised over individuals by or through the State. More than some others of his generation, perhaps, he understood the need for a "scientific outlook" that was more than the expression of scientific technique, one that incorporated the "moral outlook" of an earlier age and valued more than utility and the exercise of power. Yet Russell's attempt to articulate a moral outlook in science and to solve the dilemma posed by the "old savage in the new civilization" was ultimately unsuccessful. By 1938, he had invalidated the association of science and metaphysics in a way that ultimately left him no other grounds than the utilitarian exercise of power on which to establish a modern social ethic. These were the very grounds that he had earlier wanted to reject, as part of his concern over the preservation of individual freedoms in a scientific society.

Russell's failure to articulate a functional social morality other than one based on power may be attributed to his separation of facts from values, a separation of the knowledge proper to science from the values proper to ethics. Had he allowed for a kind of epistemology other than one based on "facts" (which, in science, had proven to be "logical fictions"), he might have found a way to articulate a different view of knowledge-as-experience that could have incorporated his understanding of instinct, impulse, and the "generosity of spirit" he associated with "love."[1] In considering religion primarily as a social phenomenon, and in criticizing the social and historical role of Christianity in western culture, he robbed himself of a basis on which to discuss the universal character of "values." Far from discovering the philosophical underpinnings for a "moral outlook" in science, by 1938, Russell found himself without defensible reasons for maintaining the interests of the individual against the organization of the State. Quite apart from its claims to truth, or the means by which those claims might be substantiated, religion or metaphysics could have served a more positive role in establishing a workable social ethic than Russell was prepared to admit. His refusal to accept the possible utility of metaphysical ideas in addressing practical problems led him to an understanding of the operations of power in modern society that was as instrumentalist as anything he had rejected in *The Scientific Outlook*.

Russell understood that solving the problem of "the old savage in the new civilization" required more than changes in political institutions. The kinds of oppositions he used to characterize life in the Machine Age, whether between industrial civilization and humanity, the individual and the machine, or freedom and organization, illustrate the same sort of dichotomy Russell also observed between reason and impulse. As we have already seen, the Great War revealed to Russell the extent to which a rationalist understanding of human nature was inadequate; civilization was a veneer over a more primitive nature ruled more by instinct and impulse than by reason. In the interwar period, he found it difficult to incorporate the positive attributes of impulse and instinct into his understanding of the organized, rational society that science made necessary. Impulsive, irrational behavior had fueled the Great War, and Russell feared that without changes to the perspective of "the old savage," the same behavior might fuel a Next War whose effects, due to the development of science, might be even more catastrophic. Russell's analysis of the character of industrial civilization is a reluctant acceptance of the *need* for social control, because he ultimately does not have confidence in any other means of ensuring that the old savage will develop his moral capacity in an age of science.

Thus neither the "good life" in modern society nor happiness were possible without some restriction on how other people, and especially the State, exercised power over individuals. Yet as totalitarian governments gained control of Italy, Germany, and the Soviet Union in the 1930s and revealed something of what the State could do in the Machine Age, it is not surprising Russell preferred to focus on how individuals should handle power. Because all power ultimately was wielded by individuals, no reform of social institutions could be successful without in some way changing the inherent tendencies of individual behavior. Thus Russell emphasized the role of education in developing citizens who would be morally and intellectually equipped to live in modern society. Similarly, instead of finding external ways of imposing an appropriate morality on a population, he wanted to cultivate it from within, by insisting that people concentrate on the rational pursuit of their own happiness rather than pursue the misery of others. Science, knowledge, and the impulse to acquire and wield power were all part of the civilization in which humanity had to live, but it was not until he addressed their relationship specifically in *Power: A New Social Analysis* that Russell was able to draw together the threads of a discussion that had begun in 1920 with his observations on Bolshevism.

LIBERAL IDEALS AND MORAL PROGRESS

During the interwar period, Russell attempted to apply liberal ideals about the social value of the individual and the individual's moral improvement through education to the problem posed by the need for a new social ethic in industrial civilization. His dilemma lay, however, in the effects of organization on the instruments by which that education might be effected. The State, as he had outlined in *Prospects* and *The Scientific Outlook,* would acquire the means to control and direct both family life and education in the scientific society, allowing the moral education and improvement of the individual to be determined by the political and economic interests of those who in turn controlled the State. Despite his calls for the cultivation of "large and generous desires" to enable the survival of the old savage in the new civilization, Russell could not escape the realization that those desires would be subject to manipulation by those who had the power to do so, through the apparatus of propaganda or the agencies of the State.

Throughout the interwar period, Russell emphasized the importance of education as a tool in shaping the desires and beliefs of individuals, and thereby bringing about moral progress. In *Education and the Good Life* (1926), he incorporated many of the ideas on the subject found elsewhere in his work to that point. Just as "the ideal system of education must be democratic,"[2] he said industrialism made possible the creation of a world "where everybody shall have a reasonable chance of happiness."[3] The connection between external and internal worlds was recurrent; though the scientific spirit needed to be cultivated, so too must the "internal self-discipline"[4] that helped create the "free citizen of the universe."[5] Rather than try to curb instincts for any reason (including "morality"), Russell asserted that, ideally, "education consists in the cultivation of instincts, not in their suppression."[6] The community would be improved along with the improvement of the individual,[7] and only then would education be serving the end of true human happiness.

Yet the problem of how power was exercised in the scientific society persisted. Individual moral progress might be a "good" thing, as far as Russell was concerned, but the State alone had the means to bring it about and therefore could define just what it meant. Whatever its theoretical appeal, nineteenth-century liberalism had been rendered archaic by the advent of the post–War scientific society, something that Russell had realized earlier in *Prospects*. While he dabbled in the modern adaptation of a liberal emphasis on the individual, none of Russell's interwar publications dealt satisfactorily with how to extend improvements in the personal morality of the old savage to the new civilization as a whole.

In *Marriage and Morals*, for example, Russell noted only two contemporary ways of characterizing society—the Marxian analysis, centering on economics, and the Freudian analysis centering on sexuality and the family relationship. Not happy with the narrowness of either approach, he stated that a new sexual ethic must first eliminate existing superstitions and second, must take into account the "entirely new factors" that are a product of modern society. He concluded that the family, and the raising of children, constituted the "only rational basis for limitations of sexual freedom,"[8] and that, for now, the patriarchal model of the family was still important.[9] Although "sex cannot dispense with an ethic," Russell said it could and should dispense with one "based solely upon ancient prohibitions propounded by uneducated people in a society totally unlike our own."[10] The difference between the old "Puritan" morality and "newer morality" was the belief that "instinct should be trained rather than thwarted"[11] from childhood on, and that self-control should not be regarded as an end in itself. The way this was to be accomplished, of course, was through education, the function of which was "to guide instinct into the directions in which it will develop useful rather than harmful activities."[12] Self-control would be applied "more to abstaining from interference with the freedom of others than to restraining one's own freedom."[13] Yet it was the State that was in charge of education and that assumed a parental role that would encourage the further development of nationalism, something Russell considered anathema in the modern age.

Despite such liberal sentiments about morality and education, Russell was drawn into considering how the operations of power affected every aspect of life in the scientific society. He observed that "power, sex, and parenthood appear to me to be the source of most things that human beings do, apart from what is necessary for self-preservation. Of these three, power begins first and ends last."[14] On the relationship between science and knowledge, he commented: "If knowledge is power, then the love of knowledge is the love of power."[15] Despite the pervasiveness of "power," however, it was not by itself to blame it for unhappiness: "Love of power . . . is a strong element in normal human nature, and as such is accepted; it becomes deplorable only when it is excessive or associated with an insufficient sense of reality."[16] Instead, ordinary "day-to-day" unhappiness resulted from "mistaken views of the world, mistaken ethics, mistaken habits of life, leading to destruction of that natural zest and appetite for possible things upon which all happiness, whether of men of animals, ultimately depends."[17] In calling his book *The Conquest of Happiness,* however, Russell indicated the preeminent importance of power in the modern world: it was not the acquisition of happiness, nor its attainment, but the *conquest* of happiness with which Russell and his audience were concerned. Some means of dealing with the pervasive problem of the exercise of power in a world dominated by scientific technique had to be found if individual happiness was to be more than an illusion, but Russell was unable to extend his prescriptions for personal morality in a plausible way to society itself.

POWER: A NEW SOCIAL ANALYSIS

Russell's published work in the interwar period reflected a gradual articulation of the principle of "power" in society and of how a scientific and industrial society had opened different avenues for its expression. *Power: A New Social Analysis* (1938) made explicit the implications of several previous references to the function of "power" in modern society. In *The Practice and Theory of Bolshevism* (1920), Russell had reflected on the abuse of power.[18] There also were two chapters in *Prospects* that dealt with the sources and distribution of power, where he had defined power as the "ability to cause other people to act as we wish" and claimed public opinion was the deciding factor in who would hold power in the future.[19] An important theme in *Icarus* was how the power yielded by science had to be wisely applied.[20]

The dominant theme of his 1938 book was the function of "power" as a principle within human nature and society. Defining it as "the production of intended effects" and a "quantitative concept,"[21] power for Russell was the fundamental concept in social science, just as energy was in physics.[22] He claimed it was impossible to isolate one form of power from others (the failing of Marx, he said, who focused primarily on economics) and said that "the laws of social dynamics are . . . only capable of being stated in terms of power in its various forms."[23] In modern society, Russell observed that power over men had become possible through power over matter[24] and tended "to generate a new mentality" with global implications:

> In former days, men sold themselves to the Devil to acquire magical powers. Nowadays they acquire these powers from science, and find themselves compelled to become devils. There is no hope for the world unless power can be tamed, and brought into the service, not of

this or that group of fanatical tyrants, but of the whole human race, white and yellow and black, fascist and communist and democrat; for science has made it inevitable that all must live or all must die.[25]

While it was necessary to tame the power expressed through scientific technique, both liberalism and Marxism (along with the other political possibilities) had failed to do so. Old-fashioned liberalism had failed because "it was only political," whereas Marxism had failed because "it was only economic."[26] He believed that only a combination of these two made a solution possible and thus by implication that a social ethic could only be expressed in political and economic terms.

As we have seen already, Russell's earlier responses to the problems posed by the exercise of power in the modern age were the typical ones of a liberal interested in "taming" it through the education and moral development of individuals, and this book continues in the same vein.[27] He was aware of the difference between what such a liberal perspective entailed for the importance of the individual in contrast to the "totalitarian" perspective that the operations of science in industrial civilization tended to create:

> The former regards the welfare of the State as resting ultimately in the welfare of the individual, while the latter regards the State as the end and individuals merely as indispensable ingredients, whose welfare must be subordinated to a mystical totality which is a cloak for the interest of the rulers.[28]

Traditional liberalism was less "powerful" than totalitarianism, an obvious conclusion given the political environment of the 1930s and the rise of fascism. Russell identified several reasons for the decline of traditional liberalism in modern society, while asserting the importance of safeguarding the rights of the individual:

> The decay of Liberalism has many causes, both technical and psychological. They are to be found in the technique of war, in the technique of production, in the increased facilities for propaganda, and in nationalism, which is itself the outcome of Liberal doctrines. . . . The problems of our time, as regards the individual to the State, are new problems, which Locke and Montesquieu will not enable us to solve.[29]

Russell also was unable to solve these problems because he could not settle on anything with which to oppose the exercise of power by the State, or those who controlled the apparatus of the scientific society.

In trying to depict "the laws of social dynamics" in terms of power, Russell was attempting an analysis of industrial civilization that went beyond the more limited analysis provided by nineteenth-century political and economic theory. No sociologist himself, he was groping for an understanding of how power relationships functioned within society and how they might be directed in a way that made the scientific society habitable by individuals, making some important observations about the social expression of power in the Machine Age. He noted the necessity of a creed for social cohesion[30] and the inevitability of power increasing with an increase in organization.[31] While religious institutions had wielded power in the past, he recognized much of that power had been transferred to the State, implying that the State had assumed some of the metaphysical importance previously assigned to the Church as well.[32] Though there were beneficial effects of increased organization, he worried about what too much organization would mean for the individual, and thus still approved of a radical or an impulsive element in society.[33]

While some recent scholarship has pointed out the overlooked merits of *Power: A New Social Analysis*,[34] Russell was disappointed by its reception.[35] In a review published in 1939, Frank H. Knight began by calling it "not only brilliantly written, but a fair proportion of the keen and witty observations which it contains have the merit of being profoundly true and practically relevant" and also "important" in connection to "saving modern civilization."[36] Having said this, Knight excoriated Russell for promoting ideas about power which he, as a philosopher, presumably should have known better than to do. Power is more than "the production of intended effects"—he observed one may possess power and yet not use it.[37] Calling the term "undefinable," Knight moved on to query the role that choice, or free will, plays, noting that "intrinsically, one configuration of matter in space has no more value, or interest of any sort, than another." Thus, he says, "The relation between value and configuration is a primary element in the problem of power."[38] This conspires to make the problem of power ("from the standpoint of social analysis") not an economic or a political problem but *"an ethical problem."*[39] Power is, for Knight, a means to an end, and the one between means and ends is fundamentally an ethical one, requiring choices between good and evil, however they may be distinguished. Knight goes on to connect this book to the final chapter in *Religion and Science* ("Science and Ethics"), observing that Russell seems to think "all persuasion is purely an exercise of power"[40] and concluding that the book's argument is ultimately "untenable" because of Russell's invalid distinction between "facts" and "values." He wryly commented that, "the impossibility of adhering to a categorical distinction between facts and values could easily be documented at virtually any length from the book under consideration,"[41] and he chastised Russell for attempting to distinguish between the "facts" of science and the "values" of ethics:

> If value judgments do not state fact, in the general sense of having validity, science evaporates into meaningless verbiage, or noise, along with ethics. One cannot in logical consistency be a moral skeptic without being an intellectual skeptic also; the same reasons that impel to one position impel equally, in principle, to the other.[42]

Knight found the final chapters of *Power* the most provoking on this point. In them, Russell grappled with the nature of the relationship between power, morality, and ethics, and he questioned on what moral basis, or according to what view of ethics, the power of the State or of individuals could be legitimately controlled, directed, or restricted. Distinguishing between "positive morality," or that associated with institutions or legal codes, and "personal morality," related to the individual or to "conscience,"[43] Russell was, in effect, distinguishing between social and personal ethics. He observed that the purpose of a traditional morality was "to make the existing social system work," and he wanted to discover whether there might be some other basis for positive morality than the exercise of power.[44] He saw the place of "personal morality" to be in the fostering of alternatives to traditional morality, painting a favorable self-portrait of an individual who was a "constructive rebel," working for the impersonal goal of a different kind of community and trying to persuade others to share his or her desires.[45]

Russell could not extricate himself, however, from the difficulties posed for social ethics by a completely subjective, individualist morality. Severing the metaphysical connection from his discussion of "happiness," for example, meant the promotion of an individualistic concept of love, in which the desires of the individual, not the group, were paramount. While he attempted to avoid this problem by calling for an ethic in which desires were harmonized rather than considered in conflict, desires of the "group" were more easily manipulated by power interests than those of the individual. While "good" and "bad" were aspects of both ethics and morality, Russell resisted the idea that either was an expression of any intrinsic value. As he put it, "there are no facts of ethics."[46] "Good" was what we desired; "bad" was what we did not desire. In terms of conduct, judgments of good or bad related to the consequences of actions, as to whether those consequences were desirable or not. Conflicts did not arise between intrinsic values of good and bad, he said, but as a result of incompatible desires.[47] "Good" then comes to characterize the collective desires of a group or a society; "doing good" may then be understood as acting in such a way as to harmonize collective desires.[48]

Yet there was ultimately no adequate logical reason for one kind of community to be preferred to another, given the dissociation Russell had already asserted between religion or metaphysics, on the one hand, and social ethics, on the other hand. That choice depended instead upon the exercise of power, manifested either as propaganda or coercion, not upon persuasion resulting from the apprehension of truth or the acquisition of knowledge. Russell was uncomfortable with such a conclusion, and he tried to find a way around it by urging the pursuit of power as a means, not as an end, and proclaiming that "the ultimate end of those who have power . . . should be to promote social cooperation, not in one group as against another, but in the whole human race."[49] Unfortunately, these sentiments could easily be dismissed as reflecting Russell's own personal desires and not some fundamental truth about reality, or a basic need of humanity. While he said that "religion and morality" were useful in combating the "feelings of unfriendliness and desire for superiority" that opposed such cooperation,[50] that recognition carried little weight in a society where he said universal values could not exist. Without that universal character, moreover, religion and morality (in whatever form) could only reflect the desires of a group of individuals, neither to be preferred nor rejected on any rational grounds over the expression of any other group.

It is ironic that Russell recognized the importance of creeds in establishing social cohesion, even though he apparently could accept no other source of these creeds apart from some shared, communal desire.[51] Apart from something that at least *functioned* as a metaphysic, whether or not it was *true,* there could be no basis for a social ethic in the scientific society other than one based upon power, expressed in terms of the instrumental operations of science and applied on a utilitarian basis for the benefit not of the individual but of society as a whole. Ultimately, Russell believed that such a power ethic would not even be satisfactory to the scientific society: "If social life is to satisfy social desires, it must be based upon some philosophy not derived from the love of power."[52]

What was clear to Russell, at least in theory, was the need for attitudes to change, if any change was to occur in social morality, because ethics for him dealt with emotions and desires rather than facts.[53] Yet the fact/value dichotomy that we observed in relation to Russell's defense of the conflict scenario in the end constituted the single greatest barrier to bridging the gulf between personal and social morality which the dilemma of the "old savage in the new civilization" required.

IN SEARCH OF AN "OPERATIONAL METAPHYSIC"

I suggest that there is a difference, therefore, between what Russell wrote in *Prospects of Industrial Civilization* and *The Scientific Outlook* and what he wrote in *Religion and Science* and *Power: A New Social Analysis.* The first two books reflect a hostility to instrumentalism and to a society constructed on the basis of the utilitarian exercise of power that resulted from the consideration of science only in terms of technique. Russell's "moral outlook in science," while it lacked a specific definition, included humanistic and metaphysical values and concerns that went beyond a narrow view of science-as-technique. By 1935, however, he had been forced to retreat from this position because of the support it implied for the metaphysical speculations of some of his contemporaries. Religion, for Russell, was not a source of knowledge; personal, religious experience was emotional, and moral values were an expression of individual desire rather than of some intrinsic verity or cosmic purpose. The social ethic required for the survival of the old savage in the new civilization could not therefore be expressed in religious or metaphysical terms. The second pair of books reflect his rejection of religion or metaphysics as a source of a social ethic and focus instead on the association between ethics and power and the need to direct the future course of society in political and economic terms. In *Freedom and Organization* (1934), Russell had concluded that "economic nationalism" had become "the dominant force in the modern world" and continued to be "the prevalent creed."[54] He maintained that, "it is not by pacifist sentiment, but by worldwide economic organization, that civilized mankind is to be saved from collective suicide."[55] Consequently, the humanistic and metaphysical concerns earlier associated with a "moral outlook in science" were set aside by 1935 in favor of the pragmatic concerns presented by the effects of power in the Machine Age.

Russell's ready acceptance of a conflictual model of the relationship between science and religion thus entailed a rejection of the one basis on which he could have constructed an ethic in opposition to that of the instrumentalists. Without at least an "operational metaphysic" to which a group of individuals could assent, Russell found himself with no way of making the leap from personal to social morality that a modern social ethic required. Individual moral progress might be accomplished through education or whatever means, but there was no assurance that it would ever be translated into the political or social structures of a scientific society. Further, without some metaphysical standard against which moral progress could be measured, there was no adequate means of adjudicating which values should be encouraged and which should be discouraged. Russell apparently explored the possibilities of a social morality grounded on something other than religious principles, as the outline of a popular book from around 1937 entitled "Need Morals Have a Religious Basis?" (which was never completed) seems to indicate.[56] Beginning with the statement that the "question is not whether religion is true, but whether, *if* it is not, any basis for ethics is possible,"[57] Russell seems to conclude that it is not. Noting that an effective ethic must appeal to and control both rational and non-rational motives,[58] he observed:

> We must find something which both the religious and the non-religious agree to call "moral" conduct, and then ask whether *this*

requires a religious basis. We must therefore look for some ethical system which does not appeal to religious authority, but yet is accepted by all, whether religious or not. There is no such system. But is there perhaps some basic principle as to which we could agree?[59]

Working his way through a series of moral principles, however, Russell was unable to settle on one.

In the absence of such an overarching moral principle, and by asserting that values had no universal character (only the specific character given to them by the desire of individuals), Russell had left himself only the utilitarian exercise of power as an effective means of directing the ethics of society as a whole. A religion such as Christianity could provide a basis for the expression of a social ethic consistent with its beliefs, while a group of individuals could give assent to metaphysical principles that might result in a consensual ethic. By invalidating the social function of religious or metaphysical beliefs within the scientific society, and by saying that personal religious experience could have only an emotive (and not a veridical) character, however, Russell left the individual at the mercy of whatever the State decided. Because the increase in scientific knowledge led to an increase in organization, and thus the concentration of power in the hands of those who controlled the State, an individual would have increasingly less control over his or her own life in the scientific society. Given the additional problem that the desires of those who controlled the State were just as likely to be irrational as the desires of anyone else, there was no reason to believe that the totalitarian future of Huxley's *Brave New World* could be avoided.

In the end, I suggest that Russell was in need of an "operational metaphysic" with which to oppose a social ethic based on the utilitarian exercise of power. This, in effect, was what Albert Schweitzer had suggested when he said that the ethics of society were dependent upon its worldview. The mechanistic conception of society that Russell had opposed in *Prospects* was, in its essence, a worldview; only an alternative worldview could effectively combat its implications for the future of industrial civilization. Russell tried to outline something of an alternative worldview in *The Scientific Outlook,* but in the end, he had shied away from its metaphysical implications. By the very end of *An Inquiry into Meaning and Truth* (1940), he grudgingly acknowledged the need in a linguistic sense for universals of some kind, but his work during the interwar period reflected a stubborn refusal to be otherwise convinced of their existence.[60]

By "operational metaphysic," I mean a metaphysic that *functions* as though it were true, while the question of its *actual* truth is set aside as something undemonstrable. It is impossible to prove that Schweitzer's "reverence for life," for example, is a universal truth, but it could function as though it were true in guiding the formulation of a social ethic in which all life was valued. Similarly, the ideas of "universal human rights" or "the equality of persons" function as metaphysics, without there being adequate rational grounds on which to make such assertions. If Russell had been able to formulate an operational metaphysic that addressed the problem of "the old savage in the new civilization," he would have found the alternative he lacked, in the interwar period, both to a morality based upon power and to the mechanistic conception of society.

Practically speaking, an appeal to such an operational metaphysic is the only means the "preacher" has to combat the power of the "legislator" in a scientific society, which is organized and directed according to the power generated by technique. Emotional persuasion in ethical matters by itself is insufficient to direct society as a whole, whatever influence it might have on the lives of individuals, because the preacher has only a limited ability to counteract the mechanisms of persuasion—and coercion—which are at the disposal of the State. Yet an ethic grounded in metaphysical or religious beliefs is difficult to suppress and gives individuals a sense of personal identification with a cause or with a Truth greater than the State or any of its instruments.

TOWARD THE NEXT WAR

The most powerful force in society is one in which the appeal of the preacher and the power of the legislator are combined. Russell's *Power* appeared in print just as Europe moved to the brink of war once again. In the persuasiveness of National Socialism in Germany and the lure of Fascism in Italy, there was an object lesson in the effectiveness of a metaphysic to bind people to a political and social purpose. It also was further evidence that Russell's abhorrence of creedalism of all kinds was well founded, but that must have been small comfort as his experience of the Great War came to be repeated. Reason and its children once again were no match for irrational behavior and emotion, and the cycle of "the old savage in the new civilization" continued, escalating toward the Armageddon that Russell and his contemporaries feared. The impotence of individual reason to resist the social impulses to war, fueled as they were by the passion of the preacher to persuade, and the organization and the coercive ability of the legislator to eradicate dissent, could not have been painted more starkly.

It is after the dust of this second World War settles and the psychological moment between it and the Next War is established again that Bertrand Russell returned to the same issues that shaped his thought in the interwar period. He focused on human knowledge and the means of knowing (*Human Knowledge: Its Scope and Limits*, 1948); he elaborated on earlier ideas about the relationship between the individual and industrial civilization (*Authority and the Individual*, 1949); and he considered the specific effects of science on society (*The Impact of Science on Society*, 1952). Finally, he combined these issues in his last major work, *Human Society in Ethics and Politics* (1954), published when he was eighty-two years old. After that, there were essays and memoirs but no further monographs on the themes that had so possessed his attention since his retreat from Pythagoras began during the Great War.

During the interwar period, Russell had both the desire and the means, through his popular publications, to agitate for the kinds of changes he believed necessary in the scientific society being created by science and technology. What he lacked was a subject that would enable him to unite his arguments with a compelling metaphysic, a worldview that could compel or persuade the old savage to turn away from technological self-destruction. Toward the end of his life, it was Russell's practical preaching against nuclear arms that brought about the coherent expression of the operational metaphysic he had previously been unable to find. Global survival, in opposition to the perils of technique in the nuclear age (perhaps a metaphysic equivalent to Schweitzer's "reverence for life"), embodied his conviction that living things are intended to live and to reproduce, not merely to die by accident or by design in a nuclear blast. Who made these living things, if anyone did, was immaterial to the need to make it possible for life to continue. His rhetoric and performances after 1945 thus were geared toward that end, and against the militarism expressed in the Vietnam conflict and its satellite wars elsewhere.

To the end of his days, Russell did not abandon his agnosticism toward truth, but he set aside the question of its philosophical proof for the practical purpose of ensuring the survival of future generations that might, in time, discover such proof for themselves. It is one of the ironies of Russell's life that the best evidence he believed such proof might exist came from his own willingness to work toward an end of which he had been so profoundly skeptical for so long. As a philosopher and logician, he could see the inexorable nature of the ABC of science and technology that took the old savage ever closer to Armageddon. As someone who cared deeply about people, whether they could accurately be described as old savages or not, Bertrand Russell persisted in his efforts to give future generations the opportunity to learn more about the universe and more about themselves, showing by his actions that within the human spirit—whatever its source—there is an unshakable connection between science and hope.

Notes

To simplify the references to Bertrand Russell's works, the following editions and abbreviations have been used in the notes:

The Problems of Philosophy. 1912; London: Thornton Butterworth, 1928. *[P Phil]*
Our Knowledge of the External World. Chicago: Open Court, 1914. *[OKEW]*
Principles of Social Reconstruction. 1916; London: Unwin, 1989. *[PSR]*
Political Ideals. 1917; London: Unwin, 1980. *[PI]*
Mysticism, Logic, and Other Essays. 2nd ed. 1917; rpt. London: George Allen and Unwin, 1950. *[ML]*
Proposed Roads to Freedom: Socialism, Anarchism, and Syndicalism. New York: Henry Holt, 1919. *[PRF]*
Introduction to Mathematical Philosophy. London: George Allen and Unwin, 1919. *[IMP]*
Bolshevism: Practice and Theory. New York: Harcourt, Brace and Howe, 1920. *[Bol]*
The Analysis of Mind. London: K. Paul, Trench, Trubner, 1921. *[AMind]*
The Prospects of Industrial Civilization. London: G. Allen and Unwin, 1923. *[PIC]*
The ABC of Atoms. London: K. Paul, Trench, Trubner, 1923. *[ABCA]*
Icarus or the Future of Science. London: K. Paul, Trench, Trubner, 1924. *[Icarus]*
The ABC of Relativity. New York: Harper, 1925. *[ABCRel]*
What I Believe. London: G. Allen and Unwin, 1925. *[WIB]*
Education and the Good Life. New York: Liveright, 1926. *[EGL]*
The Analysis of Matter. London: G. Allen and Unwin, 1927. *[AMatter]*
Outline of Philosophy. 1927; rpt. London: Unwin, 1986. *[Outline]*
Sceptical Essays. 1928; rpt. London: Unwin, 1991. *[SceptEss]*
Marriage and Morals. 1929; rpt. London: George Allen and Unwin, 1930. *[M&M]*
The Conquest of Happiness. New York: Liveright, 1930. *[COH]*
The Scientific Outlook. London: G. Allen and Unwin, 1931. *[SO]*
Education and the Social Order. London: George Allen and Unwin, 1932. *[ESO]*
Freedom and Organisation. London: G. Allen and Unwin, 1934. *[FO]*
Religion and Science. London: T. Butterworth, 1935. *[R&S]*

In Praise of Idleness and Other Essays. 1935; London: Unwin, 1990. *[PIdleness]*
Power: A New Social Analysis. London: G. Allen and Unwin, 1938. *[Power]*
An Inquiry into Meaning and Truth. London: George Allen and Unwin, 1940; rpt. 1951. *[Inquiry]*
The History of Western Philosophy. 1946; rpt. Unwin Hyman, 1989 *[HWP]*
Portraits from Memory and Other Essays. London: G. Allen and Unwin, 1956. *[PFM]*
Why I Am Not a Christian and Other Essays. 1957; rpt. London: Unwin, 1975. *[WNC]*
My Philosophical Development. 1959; rpt. London: Unwin, 1985. *[MPD]*
Fact and Fiction. 1961; London and New York: Routledge, 1994. *[F&F]*
Autobiography. 1967, 1968, 1969. Rpt. London: Unwin, 1989. *[AB]*

The Collected Papers of Bertrand Russell [CP]:
Volume 1: *Cambridge Essays 1888–89.* Eds. Kenneth Blackwell, Andrew Brink, Nicholas Griffin, Richard A. Rempel, and John G. Slater. London: G. Allen and Unwin, 1983.
Volume 2: *Philosophical Papers 1896–99.* Eds. Nicholas Griffin and Albert C. Lewis. London: Unwin Hyman, 1990.
Volume 4: *Foundations of Logic: 1903–05.* Ed. Alastair Urquhart. London and New York: Routledge, 1994.
Volume 6: *Logical and Philosophical Papers 1909–13.* Ed. John G. Slater. London and New York: Routledge, 1992.
Volume 8: *The Philosophy of Logical Atomism and Other Essays 1914–19.* Ed. John G. Slater. London: George Allen and Unwin, 1986.
Volume 9: *Essays on Language, Mind and Matter 1919–26.* Ed. John G. Slater. London: Unwin Hyman, 1988.
Volume 10: *A Fresh Look at Empiricism 1927–42.* Ed. John G. Slater. London and New York: Routledge, 1996.
Volume 12: *Contemplation and Action 1902–14.* Eds. Richard A. Rempel, Andrew Brink, and Margaret Moran. London: George Allen and Unwin, 1985.
Volume 13: *Prophecy and Dissent 1914–16.* Ed. Richard A. Rempel. London: Unwin Hyman, 1988.
The Selected Letters of Bertrand Russell. Volume 1: The Private Years (1884–1914). Ed. Nicholas Griffin. Harmondsworth, Middlesex: Allen Lane/ Penguin Press, 1992.

INTRODUCTION

1. Russell himself wrote an article on the subject as early as 1924 ("If We Are to Prevent the Next War," *The Century Magazine* 108 (May 1924): 3–12 [C24.23]). Numerous other examples of this attitude may be cited. The causes and outcome of "the war to end war" made it manifestly clear to

many people that another war was inevitable, given the political aftermath of the Treaty of Versailles, the race for developing modern weaponry despite attempts at disarmament, such as the Washington Conference in 1921, and the earlier failure of pacifist movements in the face of nationalism. The failure of the League of Nations to enforce peace in Abyssinia or Spain heightened fears that it was only a matter of time until the Next War, whatever form such a war might take. In *The Book of Joad: A Belligerent Autobiography* (1932; rpt. London: Faber and Faber, 1943), for example, C. E. M. Joad wrote that in the next war there would be no noncombatants and observed that "the international stage seems set for war" because of the armaments race and the fact that "the psychology of the people grows once again manifestly bellicose" (60). A variety of publications, popular and professional, discussed the tactics and weaponry of the Next War. See, for example, John Bakeless, ed., *The Origins of the Next War: A Study in the Tensions of the Modern World* (New York: Viking, 1926); J. F. C. Fuller, *Pegasus, or Problems of Transport* (London: K. Paul, Trench, Trubner, 1926) and *Atlantis, or America and the Future* (London: K. Paul, Trench, Trubner, n.d.); B. H. Liddell Hart, *Paris, or the Future of War* (New York: Dutton, 1925).
2. *PFM*, 30.
3. Ibid., 35.
4. *MPD*, 157.
5. Ibid.
6. Ibid.
7. Ibid., 156–57. Russell also identified a new interest in what he called "political theory" that emerged from the War and his imprisonment. Unlike his ventures into mathematical abstraction, he later wrote that political theory left him with a feeling of accomplishment (*AB*, 294).
8. *MPD*, 157.
9. Russell recognized the historical role played by Pythagoras in bridging the realms of mathematics and theology for the first time. Without Pythagoras, he said, the notion of an external world "revealed to the intellect but not to the senses," from which Christians proceeded to derive logical proof of the existence of God and immortality, would have been impossible (*HWP*, 49–56).
10. *PIC*, 74.
11. *AB*, 261.
12. He began one such article as follows: "It is commonly maintained by those whose income is derived from preaching that religious people are moral while irreligious people are immoral, and that everyone knows what is meant by religion and what by morality. I wish to challenge all these propositions" ("Morality and Religion," 1929, *CP* 10, #32, 234). During the interwar period, Russell was certainly someone who derived the majority of his income from an equivalent occupation.
13. Because of this focus on rhetorical performance, and Russell's oft-declared

position that anything he wrote worth reading had appeared in print, private correspondence becomes less significant in understanding Russell's concerns during the interwar period. It is this public Russell in which I am most interested, and so his published works—as opposed to his correspondence or private writings—have been considered his "voice" during the interwar period, for ultimately a preacher only performs for an audience.

14. *The Philosophy of Bertrand Russell* (New York: Tudor Publishing, 1951).
15. Ibid., 724.
16. Ibid.
17. For the initial idea of the importance of performance in Russell's work, I am indebted to Wayne C. Booth's *Modern Dogma and the Rhetoric of Assent*. Booth considered rhetorical performance a crucial means of relating Russell's ideas in a way that avoided the charge of inconsistency. He set out three distinct "roles," among which Bertrand Russell alternated: Russell I, "the genius of mathematical logic, sought certain knowledge about what he called 'matters of fact' or 'the world'"; Russell II, "the 'man of reason,' often the man of rational protest, tried to disestablish certain past beliefs and establish the more adequate beliefs taught, so he said, by science and the scientific view of man's motives and possibilities"; Russell III, "the man of action and passion, the poet and mystic, argued and fought for particular causes in the world, though with a steady nagging awareness that he could not really prove that his cause was just." Booth said that Russell's inability to settle for long on any one role was reflected in the apparent plasticity of his views. Different roles required a different script, and a different cast of characters. See Wayne C. Booth, in chapter 2 of his *Modern Dogma and the Rhetoric of Assent* (Notre Dame: University of Notre Dame Press, 1974), 46–47. I am indebted to the late Dr. Ben Meyer for drawing my attention to Booth's book.
18. It is interesting, for example, to see how many of these books appeared in paperback versions, as well as in hardcover.
19. *The Debunker and the American Parade* (June 1929): 3–16 (Girard, Kansas) [C29.19].
20. One such radio series, sponsored by the BBC, was published in both Britain and the United States as *Science and Religion: A Symposium* (London: Gerald Howe and New York: Charles Scribner's Sons, 1931).
21. Quoted in *CP* 10, 44.
22. Ibid., 45.
23. Paul Schilpp, ed. *The Philosophy of Bertrand Russell*. New York: Tudor, 1951.
24. A point that Russell also noted: "There is, I think, a psychological connection, but that is a different matter." What he meant by such a "psychological connection" is unclear, but it meant, at least, that they were both written by the same person. Ibid., n. 5.

25. Eduard C. Lindeman, "Russell's Concise Social Philosophy," in Schilpp, *The Philosophy of Bertrand Russell,* 560.
26. *The Incompatible Prophesies: Bertrand Russell on Science and Liberty* (Oakville, Ontario: Mosaic Press, 1978).
27. Nicholas Griffin, "The Tiergarten Program," in Ian Winchester and Kenneth Blackwell, eds., *Antimonies and Paradoxes: Studies in Russell's Early Philosophy* (Hamilton: McMaster University Library Press, 1989), 19–34; *AB,* 127, 727.
28. *Technics and Civilization* (New York: Harcourt, Brace and Company, 1934).

CHAPTER 1

1. Daniel J. Kevles, *The Physicists: The History of a Scientific Community in Modern America* (New York: Random House, 1979), 180.
2. Raymond Fosdick, *The Old Savage in the New Civilization.* (1928; Garden City, N.Y.: Doubleday and Doran, 1929). In the foreword, Fosdick stated that "four were commencement addresses given at Wellesley College, Colgate University, Vanderbilt University, and the University of Iowa. One was the Founders' Day address at Mount Holyoke; another was given before the Institute of Arts and Sciences of Columbia University; while the basis of still another was a convocation address at the University of Virginia." Other parts of the book had been presented in addresses at Colorado College, the University of Georgia, the University of Nebraska, and the University of Kansas (v).
3. Fosdick, *The Old Savage,* 37. Fosdick misquoted "Professor Schiller of Oxford" somewhat in saying that "the race stands dismayed at the prospect of the old savage passions running amok in the full panoply of civilization." The original quotation was more explicit about the dangers inherent in the unhindered development of applied science: "Science has exposed the paleolithic savage masquerading in modern garb to a series of physical and mental shocks which have endangered his equilibrium. It has also enormously extended his power and armed him with a variety of delicate and penetrating instruments which have often proved edge tools in his hands and which the utmost wisdom could hardly be trusted to use aright. Under these conditions the fighting instinct ceases to be an antiquated foible, like the hunting instinct, and becomes a deadly danger. No wonder the more prescient are dismayed at the prospect of the old savage passions running amok in the full panoply of civilization!" F. C. S. Schiller, *Tantalus, or the Future of Man,* Today and Tomorrow series (1924; rpt. New York: Dutton, 1925), 34–35.
4. Four of the early volumes helped set the philosophical parameters of the series: J. B. S. Haldane's *Daedalus, or Science and the Future;* Bertrand

Russell's *Icarus, or Science and the Future;* F. C. S. Schiller's *Tantalus, or the Future of Man;* and Russell's *What I Believe.*

5. Despite this gloomy depiction of the destructive tendencies of humanity, however, Schiller had prefaced his book with the observation that he was less pessimistic about the future of society than either J. B. S. Haldane or Bertrand Russell, because the evils they had portrayed were avoidable. He claimed humanity had always had the choice between alternative policies, for good or evil, thus the modern age was no different, proposing that the life of Tantalus paralleled human history in that neither human hopes nor fears were ever fully realized (*Tantalus,* 65–66). Schiller's Tantalus is trying to pluck fruit from the Tree of Knowledge, which is fed by a stream of water (the "Elixir of Life") that he is prevented from drinking because of the "debris of his former animal life."
6. Kevles, *The Physicists,* 248.
7. In *Science* 68 (September 28, 1928): 279–84, Millikan "disproved" Einstein's formulas for the derivation of atomic energy and concluded: "The energy available to [man] through the *disintegration* of radioactive, or any other, atoms may perhaps be sufficient to keep the corner peanut and popcorn man going, on a few street corners of our larger towns, for a long time yet to come, but that is all." Haldane wrote in *Daedalus* of the immense possibilities afforded by wind power instead of atomic power. "I do not much believe in the commercial possibility of induced radio-activity" (1924; rpt. New York: Dutton, 1925), 25–27.
8. In the *Saturday Evening Post* of March 4, 1933 ("Debunking Mars' Newest Toy"), Thomas R. Phillips disputed the threat of hostile airpower to the United States, saying, "The nations of the world are safe from air attack delivered across the ocean." Moreover, he said, "the next war will be fought with the weapons we now have. So a few predictions can be made. Airplanes will sink no battleships. Cavalry will still be needed and used."
9. "The Irony of Science: Can It Repair the Damage It Has Done?" *Forum* 86 (September 1931): 160–66.
10. Ibid., 160.
11. Ibid., 162.
12. Ibid.
13. *Technics and Civilization* (New York: Harcourt, Brace and Company, 1934).
14. Stuart Chase, *Men and Machines* (1929; rpt. New York: Macmillan, 1943).
15. Ibid., 282.
16. Ibid., 24–25.
17. (Chatto and Windus, 1932; rpt. London: Grafton Books, 1990).
18. A short-lived political movement in the late 1920s and early 1930s in the United States. Founded by Howard Scott, technocracy advocated the control of society by the engineers and scientists who would govern by logic.

19. The view of manufacturing based on Henry Ford's mass production assembly-line technique for building automobiles.
20. *Atlantis: America and the Future*, Today and Tomorrow series (London: K. Paul, Trench, Trubner, n.d.). In this book, Fuller describes his experiences in America, beginning with his purchase of a straw hat to fit in with the crowd on Wall Street and a friend's assessment of a large office building: "Some building, yep!" he answered, and then to impress me with its vastness, he added solemnly, "I reckon there are six hundred and fifty lavatories in it." Fuller wryly observed: "Six hundred and fifty ball-cocks under one roof, this surely is a symbol of vastness and a measurement of the necessities of life." From "this symbol, a magic panticle in its way," he proceeded to relate the uncivilized and crude nature of (American) machine culture (22–23).
21. Ibid., 27.
22. Ibid., 28.
23. Ibid., 33.
24. Ibid., 34.
25. Ibid., 26. In several other volumes of the *Today and Tomorrow* series, various authors expressed similar discomfort at the implications of life in the Machine Age. For Garet Garrett, machines had become the extension of mankind, making the ironic point that the advent of labor-saving machinery had not, in fact, decreased the workload on the average worker. Whereas John Gloag called for the idea of "craftmanship" to be applied to the work done with machines and worried over what would happen if a society based on such specialized skills should ever disintegrate, Martin S. Briggs fretted over the need to preserve the British countryside to prevent its becoming a mirror image of the industrial decay of America. Garet Garrett, *Ouroboros, or the Mechanical Extension of Mankind* (London: Kegan Paul, Trench, Trubner, n.d.); John Gloag, *Artifex, or the Future of Craftsmanship* (London: K. Paul, Trench, Trubner, n.d.); Martin S. Briggs, *Rusticus, or the Future of the Countryside* (London: K. Paul, Trench, Trubner, n.d.).
26. New York and London: Charles Scribner's Sons, 1930. Pupin also wrote the preface to the American edition of *Science and Religion: A Symposium* (1931).
27. Ibid., 1.
28. Ibid., 2.
29. Ibid., 28–29.
30. Ibid., 53. Pupin was not alone in these thoughts. E. E. Fournier D'Albe saw the "soul of the machine" as the extension of the human soul, just as machinery had become the extension of human limbs and intellectual capacity: "Every machine has a psychical element, a purpose, a 'soul.'" *Quo Vadimus; Glimpses of the Future* (New York: Dutton, 1925), 50. In *Automaton, or the Future of Mechanical Man* (London: K. Paul, Trench, Trubner, 1925),

H. Stafford Hatfield portrayed this idea even more starkly, as he considered "the future of the mechanical man" and enthusiastically discussed the possibilities of making automatons.

31. Fosdick, *The Old Savage*, vi.
32. Ibid., 21.
33. "This, then, is the problem: science will not wait for man to catch up. It does not hold itself responsible for the morals or the capacities of its human employers. It gives us a fire engine with which to throw water to extinguish a fire; if we want to use the engine to throw kerosene on the fire, it is our lookout. The engine is adapted to both purposes" (ibid., 25).
34. While there were other contemporary authors who wrote on similar themes, Fosdick seems most to have been influenced by Russell. Much of what Fosdick wrote was derivative of Russell's work to that point; what Fosdick did not quote directly from *Prospects of Industrial Civilization* or *Principles of Social Reconstruction*, he paraphrased throughout his book. Fosdick, *The Old Savage* 33, 50, 82. Fosdick also read *PSR* (*The Old Savage*, 15).
35. Daryl Revoldt, *Raymond B. Fosdick: Reform, Internationalism, and the Rockefeller Foundation* (Ph.D. dissertation, University of Akron, 1982), 159–201. Fosdick resigned on January 19, 1920, after the U.S. Senate refused to ratify the Treaty of Versailles and thus to join the League of Nations.
36. Ibid., 40.
37. Harry Emerson Fosdick was a well-known preacher, radio personality, and author preaching from Riverside Church in New York.
38. See Raymond Fosdick's *The Story of the Rockefeller Foundation* (New York: Harper, 1952) and his *Chronicle of a Generation: An Autobiography* (New York: Harper & Brothers, 1958).
39. Fosdick, *The Old Savage*, 64.
40. Ibid., 65.
41. Ibid.
42. Ibid., 67.
43. Ibid., 69.
44. "No matter into what remote region he may travel, an American can scarcely get away from his own civilization. Even in out-of-the-way villages where the language is unfamiliar and the roofs are still thatched it follows him like a spectre, screaming of sewing machines, type-writers, collars, canned soups, cosmetics, and the products of five-and-ten cent stores" (ibid., 70–71).
45. Ibid., 71.
46. Ibid.
47. Ibid., 73.
48. Ibid., 83.

49. Ibid. He also quoted Russell's *PIC* to this effect.
50. Ibid., 88–89.
51. Ibid., 89.
52. Ibid., 89–90.
53. Ibid., 90.
54. Robert A. Millikan, "The Alleged Sins of Science" in *Science and the New Civilization* (New York: Charles Scribner's Sons, 1930), 53.
55. Ibid., 62.
56. Ibid., 65–69.
57. Millikan, "Science and Modern Life," *Science and the New Civilization*, 11.
58. Ibid., 7.
59. Ibid., 8.
60. Ibid., 13.
61. Ibid., 14.
62. Ibid., 25.
63. Millikan, "The Alleged Sins of Science," *Science and the New Civilization*, 71.
64. Ibid., 77ff.
65. (London: Kegan Paul, Trench, Trubner and Co., 1931), v. McDougall also contributed *Janus or the Future of War* to the *Today and Tomorrow* series.
66. Ibid.
67. Ibid., 1.
68. Ibid., 2.
69. Ibid., 30.
70. "It is agreed that a modern intelligent man, conscious of his responsibilities as an inhabitant of the twentieth century, should be familiar with "the scientific outlook," he said, "but to acquire this outlook by brooding over the teachings and implications of modern physics is not easy." J. W. N. Sullivan, *Gallio or the Tyranny of Science* (London: K. Paul, Trench, Trubner, n.d.), 15.
71. "The methods found so successful in physics are applied to everything under the sun. It is pretty obvious that this is not due to some mystic, Pythagorean conviction that number is the principle of all things, but merely to mental inertia.... It is amazing the number of dull, unimaginative people who find a congenial life work in prosecuting researches in pseudo-science" (ibid., 53–54).
72. Ibid., 18–19. He disputes conclusions I. A. Richards reached in his *Science and Poetry, The New Science Series*, vol. 2. (New York: W. W. Norton, 1926).
73. Ibid., 19.
74. Ibid. The subsequent book was *The Limitations of Science* (1933; rpt. Harmondsworth, Middlesex: Penguin, 1938).
75. Ibid., 16.

76. Ibid., 17.
77. Ibid., 48–49.
78. Ibid., 30–31. He goes on to characterize the pre-Einsteinian view of the universe, and what has changed, especially according to what Arthur Eddington has written to date (1925).
79. Ibid., 78–79.
80. Ibid., 92.
81. "The old outlook did not regard values as inherent in reality. They were merely expressive of the accidental human constitution, but had no cosmic significance" (ibid.).
82. Ibid., 92–93.

CHAPTER 2

1. This interest in axioms was reflected in the titles and contents of his books: *The Foundations of Geometry; Principles of Mathematics; Our Knowledge of the External World*. After the Great War, he continued in this vein, writing *The ABC of Atoms* and *The ABC of Relativity*.
2. *FO*, 509.
3. Ibid.
4. *PIC*, 7.
5. Dora Russell was the co-author of *Prospects of Industrial Civilization*. Russell said in the preface that the book was "so much a product of mutual discussion that the ideas contained in it can scarcely be separately assigned" (*PIC*, v).
6. *PIC*, preface, iv.
7. Ibid.
8. Ibid., 164.
9. Ibid., 273.
10. Ibid.
11. Ibid., 75.
12. Ibid., 273. The idea that the mechanistic outlook was characteristic of society in the Machine Age was not an original observation. See, for example, Thorstein Veblen, "The Place of Science in Modern Civilization," in *The Place of Science in Modern Civilization and Other Essays* (New York: B. W. Huebsch, 1919), 29–30.
13. Even though it has "a framework and a formula," he only "discovered" them when he had written "all except the first and last words" (*AB*, 242).
14. Ibid., 242.
15. "The civilized world has need of fundamental change if it is to be saved from decay . . . But until the war is ended, there is little use in detail, since we do not know the kind of world the war will leave. The only thing that

seems indubitable is that much new thought will be required in the new world produced by the war" (PSR, 167).
16. Ibid., 157.
17. Ibid., 158.
18. PIC, preface, v.
19. Ibid., 8–9.
20. Ibid., 9.
21. Ibid., 10.
22. Ibid., 25.
23. Ibid., 16.
24. Ibid., 19. "A person afflicted with nationalism believes that his own country is the most civilized and humane country in the world, while its enemies are guilty of every imaginable atrocity and vileness. Since they are so vile and atrocious, while we are so civilized and humane, there is no degree of vileness and atrocity which we may not legitimately practice against them" (ibid., 17).
25. "The destructiveness of war being great, the fear of it is greater; therefore the intensity of national feeling is greater; and therefore the likelihood of war is greater. It may therefore be laid down as a general proposition that whatever increases the harmfulness of war also increases its likelihood" (ibid., 65).
26. Ibid., 57.
27. Ibid., 77.
28. This was because of the self-sufficiency of its North American position in relation to the rest of the world, its "large white population," and the fact that "the Americans surpass even the British in sagacity, apparent moderation, and the skillfull use of a hypocrisy by which even they themselves are deceived." Against such a formidable combination of advantages, he said, no other state could hope to prove victorious (ibid., 82).
29. Ibid., 83.
30. Ibid., 84.
31. "The claim to complete national independence on the part of every group which happens to have the sentiment of nationality is quite incompatible with the continued existence of an ordered society ... The rights of a nation as against humanity are no more absolute than the rights of an individual as against the community" (ibid., 96–97).
32. Ibid., 97.
33. Ibid., 42–43.
34. As Russell put it, in describing a letter to New York newspapers from A. C. Bedford, president of Standard Oil (on November 25, 1920), which discussed the export of Italian workers, "We have traveled far from liberal ideas when labor can be treated as a commodity of export to be exchanged against coal" (ibid., 52).
35. Ibid., 49.

36. Ibid., 54–55.
37. Ibid., 55.
38. "Therefore, when we are considering the prospects of industrial civilization, a universal class war must be regarded as a dead end, not as the fiery gateway to a new world" (ibid., 126).
39. *Bol*, 149, 167.
40. Ibid., 84–85.
41. Ibid.
42. While tending to use the word as vaguely in his own work as he noted it was used elsewhere, Russell did define socialism in this way: in terms of economics, it was concerned with the production and distribution of goods. All land and capital must be the property of the state, and what is paid for each kind of work must be fixed by a public authority, "with a minimum of what is required for bare necessities, and a maximum of what will give the greatest incentive to efficient work." There was, however, "no need of equality of income for all as part of the definition of socialism." In terms of politics, socialism required "that all sane adults should have an equal share of ultimate political power" (*PIC*, 98–99).
43. *PI*, 20. Russell's association with liberalism is not easily depicted, as Michael Freeden observed: "It is difficult to categorize Russell, who shared many of the central liberal assumptions about human reason, the dangers of the centralized state, and the importance of the individual, but in other senses remained committed to forms of social organization and economic distribution unacceptable to most liberals, and was only marginally associated with the thinkers examined in this study. In 'Socialism and Liberal Ideals' (*English Review* 30 [1920]: 339–55, 499–508), he announced his move away from liberalism to socialism, because those liberal ideals to which he subscribed could not be achieved without transforming the economic structure of society. He is, however, a further example of the flexibility of the boundaries of liberalism . . . and his books were warmly, though not uncritically, welcomed by liberal progressives" (*Liberalism Divided: A Study in British Political Thought* (Oxford: Clarendon, 1986), 68, n. 96).
44. A variation of syndicalism, in which the industrial workers took control of the industries in which they worked and ran them for their own benefit under the general direction of a socialist government in which there was no private ownership of the means of production.
45. *PRF*, 186–209.
46. *Bol*, 167.
47. *PIC*, 5.
48. *Icarus*, 5.
49. Ibid., 57.
50. *SO*, 11.
51. Ibid.

52. Ibid., 12.
53. Ibid.
54. Ibid., 274.
55. Ibid.
56. Ibid.
57. Ibid., 149.
58. Ibid., 209.
59. Ibid., 219.
60. Ibid., 220.
61. Ibid., 221.
62. Ibid., 221–22.
63. Ibid., 226.
64. Ibid., 241.
65. Ibid.
66. Ibid., 263.
67. Ibid., 268.
68. Ibid., 242.
69. Ibid., 268.
70. Ibid., 269.
71. Ibid.
72. *Political Ideals* was banned from publication in Great Britain, but it appeared in the United States in 1917.
73. *PI*, 9–13.
74. *PI*, 45.
75. *PIC*, 38.
76. Ibid., 39. Russell tried to assign blame for this "utilitarianizing of men's outlook" not to industrialism but to the fact that its growth had been dominated by "commercialism and competition." Consequently, he felt that a socialistic industrialism would yield benefits to society not obtainable through capitalism. This was not a convincing point, for there was no particular reason to think that such a change would take place in Western society. Russell himself recognized the similarities between capitalism and state socialism, as far as industrialism was concerned.
77. Ibid., 40.
78. *PIC*, 39.
79. *SO*, 40–41.
80. Ibid., 270.
81. Ibid.
82. Russell's opposition to pragmatism was not a product of the interwar period. He corresponded extensively with chief British proponent F. C. S. Schiller, and wrote reviews both of Schiller's books and those of William James. See *CP* 6, #s 21–25, and Appendix 1, for example.
83. *SO*, 273. The blame for this development Russell laid at the feet of modern

physics, which had stripped away the knowledge humanity thought it had of the external world and replaced it with "a skeleton of rattling bones, cold and dreadful, perhaps a mere phantasm," a "desert" revealed by the formulae of physicists "appalled" at their handiwork.

84. Ibid.
85. Ibid., 273–74.
86. Ibid., 274. Except, he noted, "ascetic renunciation."
87. Ibid., 274–75.
88. Ibid., 275.
89. Ibid.
90. Ibid.
91. Ibid., 278.
92. Ibid.
93. Ibid., 279.
94. Ibid., 277–78.
95. Ibid., 279.
96. Ibid.
97. *The Philosophy of Civilization*. Trans. C. T. Campion (New York: Macmillan, 1949). The translator makes a great deal of the translation of *Weltanschaaung* as "worldview."
98. CP 9, #60, 351–52.
99. "Speaking causally, our ethics are an effect of our actions, not vice versa; instead of practicing what we preach, we find it more convenient to preach what we practice. When our practice leads us to disasters we tend to alter it, and at the same time to alter our ethics; but the alteration of our ethics is not the cause of the alteration of our practice" (ibid., 353).
100. Schweitzer, *The Philosophy of Civilization*, 49.
101. Ibid., 229.
102. Ibid.

CHAPTER 3

1. (New York: Ray Long and Richard Smith, 1932), 3.
2. Ibid.
3. Ibid., 4.
4. Ibid., 5.
5. Ibid.
6. Ibid.
7. Ibid., 6.
8. AB, 443–45. The final list of contributors, however, belies the inanity of the initial solicitation, a copy of which is found in Russell's *Autobiography*. Durant also wrote to Josef Stalin, Benito Mussolini, Thomas Edison, Winston Churchill, Igor Stravinsky, Guglielmo Marconi, and Dean Inge, among others.

9. Durant, *On the Meaning of Life,* 106.
10. *AB,* 445.
11. As early as 1893 Russell had written to Alys Pearsall that he was "utterly out of sympathy with Christianity." See Nicholas Griffin, ed., *The Selected Letters of Bertrand Russell, Volume 1: The Private Years (1884–1914)* (Harmondsworth, Middlesex: Allen Lane, Penguin Press, 1992), 30.
12. In William James's classic book, *The Varieties of Religious Experience* (1902; rpt. London: Collins, 1960), James compared religious experiences and separated their common elements in a way that made such experiences at least a product of human nature, if not of some supernatural influence. His book accorded well with the development of the scientific study of religion, as scholars considered whether the origins of religious experience might be found in something other than revelation.
13. In Stefan Andersson's recent work on Russell and religion (*In Quest of Certainty: Bertrand Russell's search for certainty in religion and mathematics up to The Principles of Mathematics* (1903), Ph.D. dissertation (Stockholm: Almqvist & Wiksell, 1994), "Religion in the Russell Family," *Russell* n.s. 13, no. 2 (winter 1993–1994): 117–49 and (with Louis Greenspan) the introduction and commentary to *Russell on Religion* (London and New York: Routledge, 1999), and Nicholas Griffin's "Bertrand Russell as a Critic of Religion," *Studies in Religion* 24: 1 (1995): 47–58, the primary focus is on Russell's perspectives on religion before the Great War. I maintain that the social and ethical dimensions of Russell's retreat from Pythagoras render problematic the extension of his prior religiosity (or lack of it) to the interwar period.
14. *R&S,* 15.
15. Ibid.
16. Edgar Brightman, "Russell's Philosophy of Religion," in Paul Schilpp, ed., *The Philosophy of Bertrand Russell* (New York: Tudor, 1951), 539–56.
17. "Has Religion Made Useful Contributions to Civilization?" in *WNC,* 27.
18. Ibid., 27.
19. Ibid.
20. Ibid., 31.
21. Ibid., 40.
22. Ibid., 41.
23. Ibid., 42.
24. "The Free Man's Worship, " in *ML,* 48.
25. Ibid., 50.
26. Ibid.
27. Ibid.
28. See, for example, *CP* 12, 110–11. By 1911, he was relieved to understand that what Ottoline meant by God "is very much what I call infinity" (Griffin, ed., *Selected Letters,* 410).

29. "The Essence of Religion, " in *CP* 12, 113.
30. Griffin, 413–14.
31. Ibid., 414.
32. "The Essence of Religion," in *CP* 12, 114.
33. See Brightman, in Schilpp, *The Philosophy of Bertrand Russell,* and Russell's comments on Brightman (725–27).
34. "Has Religion," in *WNC,* 27.
35. In his introduction to *CP* 10, John Slater noted that, "during the First World War the bellicose behavior of most of the clergy greatly increased his contempt for organized religion, or 'the church' as he usually referred to it" (*CP* 10, xvii).
36. See, for example, "Morality and Religion" (1929), in *CP* 10, #32, 234.
37. "This lecture was delivered on March 6, 1927, at Battersea Town Hall, under the auspices of the South London Branch of the National Secular Society" (*WNC,* 13).
38. Kenneth Blackwell and Harry Ruja, *A Bibliography of Bertrand Russell, Volume 2, Serial Publications 1890–1990* (New York and London: Routledge, 1995), Plate III; John Slater, in *CP* 10, xxiii.
39. "Morality and Religion, " in *CP* 10, #32, 234.
40. *Bol,* 117.
41. Ibid.
42. Ibid., 117, 118.
43. Ibid., 6.
44. *PSR,* 139.
45. It is here that I would place Russell's public disavowals of Christianity, of the existence of God, of immortality, and so on. As A. J. Ayer put it, "[Russell] does not find the proposition that such a being [God] exists unintelligible, or logically impossible: he maintains only that there is not the slightest reason to think it true." Russell does not advance proofs for the nonexistence of God. He states instead that there is no means of determining the truth or falsity of claims either for or against the existence of God, the Buddha, or whomever. See Ayer, *Bertrand Russell* (New York: Viking, 1972), 131.
46. See, for example, *Outline,* 180.
47. *WNC,* 20–21.
48. Ibid., 21–22.
49. *Thrasymachus, or the Future of Morals* (New York: Dutton, 1926), 87.
50. Ibid., 86–88.
51. Plato, *The Republic,* 2nd. ed. trans. Desmond Lee. (Harmondsworth, Middlesex: Penguin, 1977), 77ff.
52. *WIB,* 5.
53. In the lead essay of *Christianity and the Crisis,* ed. Percy Dearmer (London: Victor Gollancz, 1933), E. A. Burroughs, the Bishop of Ripon, cites *The*

Conquest of Happiness as evidence of Russell's desire "in effect, for a new divine humanity without God as the means of producing it" (25). Russell's "eclectic pessimism" consists "in pointing out a way which neither he nor humanity can follow" (20, 23–25).

54. *Christian Century* XLI (September 3, 1925): 1093–95.
55. WIB, 13.
56. Ibid., 17.
57. Ibid., 19.
58. Ibid., 24.
59. Ibid., 40.
60. Ibid., 28.
61. Ibid., 32.
62. Ibid., 37.
63. Ibid.
64. Ibid., 43–44.
65. Ibid., 63ff.
66. Ibid., 72.
67. Ibid., 95.
68. It was Russell's first foray into human psychology, but however innovative his comments were regarding the importance and character of human impulses at the time, they were quickly overshadowed, even in his own writing, by developments in the young science of psychology. Throughout the interwar period, Russell referred admiringly to James Watson, Pavlov, and other pioneers in the area of behavioral psychology. While he acknowledged the rudimentary nature of what had so far been discovered, and the likelihood that psychology would never achieve the same clarity in expression as modern physics, Russell regarded it and the other social sciences as essential disciplines.
69. See the discussion in Chapter 5 on *Religion and Science*.
70. ML, 1.
71. Ibid.
72. Ibid., 8.
73. Ibid., 9.
74. Ibid.
75. Ibid., 10.
76. Ibid.
77. Ibid., 11.
78. Ibid.
79. Ibid.
80. Ibid., 12.
81. Ibid.
82. Ibid.
83. Ibid., 12–13.

84. Ibid., 13.
85. Ibid., 18.
86. Ibid., 32.

CHAPTER 4

1. In "Worlds of Physics and of Sense," Russell complained that, "physicists, ignorant and contemptuous of philosophy, have been content to assume their particles, points, and instants in practice, while conceding, with ironical politeness, that their concepts laid no claim to metaphysical validity." In a note to the revised edition in 1926, Russell wrote: "This was written in 1914. Since then, largely as a result of the general theory of relativity, a great deal of valuable work has been done; I should wish to mention specially Professor Eddington, Dr. Whitehead, and Dr. Broad as having contributed, from different angles, to the solution of the problems dealt with in this lecture" (*OKEW*, 130–31).
2. "A Mathematician Looks at Science," in *Science and Human Life* (New York and London: Harper and Brothers, 1933), 251–52. Further, he criticized Russell, as well as Eddington and Jeans, for dealing too much with mathematics and symbols and not enough with experiments and the processes of life.
3. By "new physics," I mean the discoveries associated with quantum mechanics and relativity theory, rooted in the work of Albert Einstein, but also those ideas developed by Paul Dirac, Werner Heisenberg, Nils Bohr, and others. Obviously the specific picture of what the new physics entailed developed as well, from the publication of Einstein's papers on special relativity in 1905 and general relativity in 1915 to later work by Heisenberg and Bohr, for example.
4. *Outline*, 77.
5. Ibid., 125.
6. As he wrote in the introduction to his *Sceptical Essays,* "I wish to propose . . . a doctrine which may, I fear, appear wildly paradoxical and subversive. The doctrine in question is this: that it is undesirable to believe a proposition when there is no ground whatever for supposing it true" (*SceptEss,* 11).
7. "The skepticism that I advocate amounts only to this: (1) that when the experts are agreed, the opposite opinion cannot be held to be certain; (2) that when they are not agreed, no opinion can be regarded as certain by a non-expert; and (3) that when they all hold that no sufficient grounds for a positive opinion exist, the ordinary man would do well to suspend his judgment" (ibid., 11–12).
8. *Outline,* 84.
9. "So long as we continue to think in terms of bodies moving, and try to adjust

this way of thinking to the new ideas by successive corrections, we shall only get more and more confused. The only way to get clear is to make a fresh start, with events instead of bodies" (ibid., 87).
10. *Outline*, 125.
11. *ABCRel*, 208.
12. *Outline*, 84.
13. Ibid., 83.
14. "What do we mean by a 'piece of matter' in this statement? We do not mean something that preserves a simple identity throughout its history, nor do we mean something hard and solid, nor even a hypothetical thing-in-itself known only through its effects. We mean the 'effects' themselves, only that we no longer invoke an unknowable cause for them . . . . All this, however, is only a convenient way of describing what happens elsewhere, namely the radiation of energy away from the center. As to what goes on in the center itself, if anything, physics is silent" (ibid., 126).
15. *ABCA*, 151.
16. "For philosophy, far the most important thing about the theory of relativity is the abolition of the one cosmic time and the one persistent space, and the substitution of space-time in place of both. This is a change of quite enormous importance, because it alters fundamentally our notion of the structure of the physical world, and has, I think, repercussions in psychology" (*Outline*, 86).
17. "Commonsense and pre-relativity physicists believed that, if two events happen in different places, there must always be a definite answer, in theory, to the question whether they were simultaneous. This is found to be a mistake" (ibid., 86).
18. *ABCRel*, 208.
19. Ibid., 209.
20. Ibid.
21. Ibid., 215.
22. "There is merely an observed law of succession from next to next. . . . A string of events connected, in this way, by an approximate intrinsic law of development is called one piece of matter. This is what I mean by saying that the unity of a piece of matter is causal" (*Outline*, 89).
23. Ibid., 93.
24. Ibid., 114.
25. Ibid., 109.
26. Ibid., 107.
27. Ibid., 110.
28. Ibid., 118.
29. "The knowledge we derive from physics is so abstract that we are not warranted in saying that what goes on in the physical world is, or is not, intrinsically very different from the events that we know through our own experience" (ibid., 113).

30. "There is only one definition of the words that is unobjectionable: 'physical' is what is dealt with by physics, and 'mental' is what is dealt with by psychology" (ibid., 112).
31. Ibid., 111.
32. Ibid., 117.
33. *AMind*, 307.
34. He proposed this to counter the "anti-materialistic tendencies of physics" discussed above, and the "materialistic tendencies of psychology" that linked mental activity to physical processes (ibid., 6).
35. *Outline*, 226.
36. Ibid., 239.
37. Ibid., 242.
38. Ibid., 241.
39. (1928; rpt. New York: Macmillan; Cambridge: Cambridge University Press, 1929).
40. *SO*, 15.
41. Ibid., 16.
42. Ibid.
43. Ibid.
44. Ibid., 18.
45. Ibid., 58.
46. "Although this may seem a paradox, all exact science is dominated by the idea of approximation. When a man tells you that he knows the exact truth about anything, you are safe in inferring that he is an inexact man" (ibid., 65).
47. Ibid., 69.
48. Ibid., 67–68.
49. Ibid., 75.
50. Ibid., 76–77.
51. Ibid., 78. "The metaphysic of Bergson, for example, is undoubtedly pleasant." Russell would return to this theme in *Religion and Science*.
52. Ibid., 78–79.
53. Ibid., 83.
54. Ibid., 82. See George Santayana, *Skepticism and Animal Faith: Introduction to a System of Philosophy* (1923; rpt. n.p.: Dover, 1955).
55. *SO*, 85.
56. Ibid.
57. Ibid., 86.
58. Ibid., 87.
59. Ibid., 88.
60. Ibid.
61. Ibid., 88–89.
62. In *The Analysis of Mind (AMind)*, Russell cited Eddington's *Space, Time,*

and Gravitation (5); in *The Analysis of Matter (AMatter),* he cited this book and *The Mathematical Theory of Relativity (*1925; rpt. Cambridge: Cambridge University Press, 1937) as well. He discusses *The Mathematical Theory of Relativity* in chapter 7 on "the Method of Tensors" (77–78), and in chapter 9 on "Invariants and their Physical Interpretation" (84). In a note on page 84, he also referred to Eddington's contribution to *Science, Religion and Reality* (New York: Macmillan, 1925). Eddington quoted Russell's *Introduction to Mathematical Philosophy* favorably in *Space, Time, and Gravitation: An Outline of the General Relativity Theory* (Cambridge: Cambridge University Press, 1921), 197, and later in *The Philosophy of the Physical Sciences* (1938; rpt. Ann Arbor: University of Michigan Press, 1958), 152. In *The Nature of the Physical World,* Eddington twice cited *The Analysis of Matter* (160, 278).

63. Eddington acknowledged his philosophical debt to Russell in his *New Pathways in Science* (1934; rpt. Ann Arbor: University of Michigan Press, 1959), 305–06: "I think that he more than any other writer has influenced the development of my philosophical views, and my debt to him is great indeed."
64. *The Nature of the Physical World,* 303.
65. Eddington, *The Nature of the Physical World,* 333. "From this perspective, we recognize a spiritual world alongside the physical world. Experience—that is to say, the self cum environment—comprises more than can be embraced in the physical world, restricted as it is to a complex of metrical symbols" (ibid., 288).
66. "To put the conclusion crudely—the stuff of the world is mind-stuff" (ibid., 276).
67. Ibid., 350.
68. "I repudiate the idea of proving the distinctive beliefs of religion either from the data of physical science or by the methods of physical science" (Eddington, *The Nature of the Physical World,* 333).
69. Ibid., 353.
70. Eddington, *Science and the Unseen World.* Swarthmore Lecture, 1929 (New York: Macmillan, 1929). The dust jacket of *Science and the Unseen World* reads: "Every religiously minded reader of the last four chapters of *The Nature of the Physical World* will be eager to lay hold of these additional observations of the foremost living exponent of the seen in regard to the unseen world."
71. Eddington, *The Nature of the Physical World,* 324.
72. Eddington, *Science and the Unseen World,* 47.
73. Ibid., 37.
74. Ibid., 73–74.
75. Ibid., 82.
76. "Professor Eddington is unquestionably the most delightful of all the writers on physics at the present time, and his qualities are shown almost

equally in his popular writings and in his technical mathematics. As a theoretical physicist he has a rare clarity and systematic comprehensiveness which is probably the basis of his extraordinary skill as a popularizer" (*CP* 10, #8, 53).
77. *SO,* 90–91.
78. Ibid., 94–95.
79. Ibid., 97.
80. Ibid., 96.
81. Ibid., 97.
82. Ibid.
83. "I suppose that machines will survive the collapse of science, just as parsons have survived the collapse of theology, but in the one case as in the other they will cease to be viewed with reverence and awe" (ibid., 98).
84. Russell was perturbed at the equanimity with which Eddington and others could seemingly accept the demise of causality. In his section on *The Nature of the Physical World* in *The Scientific Outlook,* he wrote: "Those who desire caprice in the physical world seem to me have failed to realize what this would involve. All inference in regard to the course of nature is causal, and if nature is not subject to causal laws all such inference must fail. We cannot, in that case, know anything outside of our personal experience; indeed, strictly speaking, we can only know our experience in the present moment, since all memory depends upon causal laws" (*SO,* 111).
85. "It requires a certain robustness to be optimistic on the basis of such a philosophy; but I suspect Eddington has secret beliefs in reserve which he has not set forth in these lectures." Russell was right, as the later publication of Eddington's *Science and the Unseen World* demonstrated (ibid.).
86. (New York: Macmillan; Cambridge: Cambridge University Press, 1930), 140.
87. Ibid., 148–49.
88. Ibid., 136.
89. Ibid., 146.
90. Ibid., 29.
91. Ibid., 140.
92. Ibid., 144.
93. *CP* 10, #12, 65.
94. "Jeans argues that the world must have been created by a mathematician for the pleasure of seeing these laws in operation. If he had ever attempted to set out this argument formally, I cannot doubt that he would have seen how fallacious it is. To begin with, it seems probable that any world, no matter what, could be brought by a mathematician of sufficient skill within the scope of general laws. If this be so, the mathematical character of modern physics is not a fact about the world, but merely a tribute to the skill of the physicist. In the second place, if God were as pure a mathematician as

His knightly champion supposes, He would have no wish to give a gross external existence to His thoughts" (*CP* 10, #12, 66).

95. "I once knew an extremely learned and orthodox theologian who told me that as the result of long study he had come to understand everything except why God created the world. I commend this puzzle to the attention of Sir James Jeans, and I hope that he will comfort the theologians and the defenders of private property by dealing with it at no distant date" (ibid.).
96. *SO*, 100.
97. Ibid., 100–101.
98. Ibid., 101. Here he criticized James Jeans's idea that God had to be some kind of mathematician; the existence of a multiplicity of geometries (and, hence, conceivable worlds) made this impossible.
99. Ibid., 136.
100. Ibid., 137–38.
101. The extent of his ire is reflected in what followed: "Every day some new physicist publishes a new pious volume to conceal from himself and others the fact that in his scientific capacity he has plunged the world into unreason and unreality" (ibid., 98–99).
102. Ibid., 99.
103. Ibid., 99–100. The exception that Russell noted was Robert Millikan, who showed a preference for natural law. Of course, Millikan went on to disgrace himself professionally by claiming to find, in the existence of cosmic rays, indisputable physical proof of the existence of God. See Kevles, *The Physicists*, 179–80.
104. *SO*, 98.
105. Ibid., 101.
106. Ibid., 102.
107. Ibid.
108. In his provision of historical examples of scientific method, and of the individuals who were responsible for discoveries of scientific "truth," Russell presented a pantheon of the "heroes of science" in *The Scientific Outlook*, to which he would later return in *Religion and Science*. Always the individualist, Russell believed that the entire course of modern civilization and the development of science would have been changed irrevocably if certain individuals had not existed, or if they had been prevented from discovering some essential truths about nature: "It is customary amongst a certain school of sociologists to minimize the importance of intelligence, and to attribute all great events to large impersonal causes. I believe this to be an entire delusion. I believe that if a hundred of the men of the seventeenth century had been killed in infancy, the modern world would not exist. And of these hundred, Galileo is the chief" (35). In that pantheon, he gave special attention to Archimedes, Galileo, Kepler, Newton, Darwin, and Pavlov. In each instance, he pointed out how the individual had discovered some fact

which there was a reason to believe was true, even if that fact was uncongenial to the wishes or beliefs of his contemporaries. Russell's own point remained clear: "Scientific method sweeps aside our wishes and endeavors to arrive at opinions in which wishes play no part" (45).
109. Ibid., 104.
110. Ibid.
111. Ibid.
112. In *New Pathways in Science,* Eddington took exception to the way he claimed that Russell misrepresented his understanding of the relationship between science and religion (306–08). In a very interesting but much less popular book, *Philosophical Aspects of Modern Science* (London: George Allen and Unwin, 1932), C. E. M. Joad tackles the misrepresentations of Eddington, Jeans, and Russell as well.

CHAPTER 5

1. Various scholars have alluded to different problems with what will be termed the "conflict scenario," notably Colin Russell, John Hedley Brooke, Geoffrey Cantor, James Moore, and David Wilson, among others. Brooke, in his *Science and Religion: Some Historical Perspectives* (Cambridge: Cambridge University Press, 1991), for example, considers the relationship between Science and Religion from the sixteenth century to the present in a way that discredits both a simplistic view of either their conflict or their reconciliation. Most recently, Brooke and Geoffrey Cantor, in the publication of their joint Gifford Lectures 1995–1996, *Reconstructing Nature: The Engagement of Science and Religion* (Edinburgh: T & T Clark, 1998), discuss a variety of historical examples that make the conflict scenario or other master narratives of the science and religion discourse untenable.
2. *Oxford English Dictionary,* 2nd. ed., prepared by J. A. Simpson and E. S. C. Weiner (Oxford: Clarendon, 1989), XIV, "scientist." Whewell wrote first in the *Quarterly Review* in 1834 that the term should now be "scientist," and later in his *The Philosophy of the Inductive Sciences* (1840). See Jack Morrell and Arnold Thackray, *Gentlemen of Science* (Oxford: Clarendon, 1981), 20, and Steven Shapin, *The Scientific Revolution* (1996; rpt. Chicago: University of Chicago Press, 1998), 5–6, n. 3.
3. David B. Wilson argues that the terms "science and religion" seriously misrepresent the historical situations in which they are used, and he suggests their replacement by a series of descriptive terms ("On the Importance of Eliminating Science and Religion from the History of Science and Religion: The Cases of Oliver Lodge, J. H. Jeans's and A. S. Eddington," in *Facets of Faith and Science,* ed. Jitse van der Meer, vol. 2 (Lanham, Md.: University Press of America, 1996). James Moore makes a similar argument, saying that the "old duality" should be replaced by a series of "covering concepts." He

suggests three: language, practice, and vocation ("Speaking of 'Science and Religion'—Then and Now," in *History of Science* XXX [1992]: 311–23).
4. For a concise account of why the conflict scenario is a recent invention, see David C. Lindberg and Ronald L. Numbers, "Beyond War and Peace: A Reappraisal of the Encounter between Christianity and Science," *Church History* 55: 3 (September 1986): 338–54. See also Brooke, *Science and Religion*, 34–42.
5. J. W. Draper, *History of the Conflict between Religion and Science* (1874; rpt. New York: D. Appleton and Company, 1897), vi.
6. Ibid., 367.
7. Ibid., 362.
8. Lindberg and Numbers,"Beyond War and Peace," 339.
9. Ibid.
10. Andrew Dickson White, *A History of the Warfare of Science with Theology in Christendom.* 2 vols. (1896; rpt. New York and London: D. Appleton, 1926), 1, ix.
11. Ibid., xii.
12. Ibid. Only then, he says, will the prophecy of Micah, the definition of St. James of "pure religion and undefiled," and "the precepts and ideals of the blessed Founder of Christianity himself, be brought to bear more and more effectively on mankind."
13. See, for example, James Moore's critique of the conflict metaphor in *The Post–Darwinian Controversies: A Study of the Protestant Struggle to Come to Terms with Darwin in Great Britain and America, 1870–1900* (Cambridge: Cambridge University Press, 1979).
14. Even prominent figures in the modern history of science have succumbed to this analysis. George Sarton, in his introduction to the incomplete reprint of *Science, Religion, and Reality* (New York: George Braziller, 1955), applauds White's view of the historical relationship between science and religion: "White's history is very deeply interesting; it is a very useful book" (14). Sarton's introduction replaces A. J. Balfour's original introduction, and Dean W. R. Inge's conclusion to the 1925 edition simply disappears without comment.
15. As recent scholarship points out, Galileo's problems stemmed as much from his personality as from his ideas; to place the Aristotelian response to heliocentrism in the mouth of Simplicius (mimicking the Pope's reaction) was an arrogance the Pope could not overlook. After Galileo's trial and abjuration of heliocentrism, he was sentenced in his old age to a house arrest sufficiently comfortable for him to produce his most important work on heliocentrism. Similarly, as James Moore and others have pointed out, the objections to Darwin's theory were based on scientific rather than theological grounds, and clergy at the time were as likely to defend him as to assail evolutionism. Only later—arguably after Draper and White set the ideological

tone of inevitable conflict between science and religion—did the hostility develop that led to the Scopes "monkey trial."
16. See, for example, E. E. Slosson, *Keeping Up With Science* (New York: Harcourt Brace, 1924) and E. E. Slosson and Otis Caldwell, *Science Remaking the World* (Garden City, N.Y.: Garden City Publishing, 1925). Slosson had a Ph.D. in chemistry and was the first editor of the Science Service, established by the American Association for the Advancement of Science, in partnership with Edward Scripps, a newspaper magnate. Another of Slosson's books, *Creative Chemistry*, sold 200,000 copies. See Kevles, *The Physicists*, 171–75.
17. Joseph Needham, ed., *Science, Religion, and Reality*. (New York: Macmillan, 1925). Needham also wrote *The Sceptical Biologist* (London: Chatto and Windus, 1929), which contained his own essays on the philosophy of biology, the relationship between science and religion, materialism and religion, and other topics in the social history of science.
18. Balfour was a Gifford lecturer, whose lectures were published as *Theism and Thought* and *Theism and Humanism*.
19. Inge was dean of St. Paul's; he also contributed to *Science and Religion: A Symposium* and was an important figure in the popular Christian press for much of the interwar period. As a Gifford lecturer, he published a two-volume work on Plotinus, which Russell reviewed favorably (*CP* 9, #63, 375–78).
20. Among the other contributors were Charles Singer, Antonio Aliotta, John Oman, William Brown, Clement C. Webb, and Needham himself (who wrote "Mechanistic Biology and the Religious Consciousness").
21. Russell referred to the difference between Eddington's views in *The Mathematical Theory of Relativity* and "The Domain of Physical Science" in *The Analysis of Matter*, 84.
22. *Science and Religion: A Symposium* (London: Gerald Howe, 1931), "Note," v.
23. The flyleaf reads, for example: "The reconciliation of science with religion may justly be described as the question of the day. The last few years have witnessed a revolutionary change in the attitude of prominent scientists towards religion, and a growing desire on the part of religious leaders to adjust their thought to scientific conclusions. Both sides are ready to admit that many of their past beliefs were ill-founded. But what has taken their place? What are we justified in believing today? It is the purpose of this symposium, recently broadcast, to answer these questions by presenting the considered views of men eminent as scientists, philosophers, and churchmen."
24. *Science and Religion: A Symposium* (New York: Charles Scribner's Sons, 1925), 4.
25. Ibid.
26. Ibid., 4.
27. Ernest William Barnes, *Scientific Theory and Religion* (New York: Macmillan, 1933), 588.

28. Ibid., 5.
29. A. J. Balfour, "Introduction," in Needham, ed., *Science, Religion, and Reality*, 5.
30. Ibid., 10.
31. Ibid., 16.
32. "Conclusion," in Needham, ed., *Science, Religion, and Reality*, 347.
33. Ibid., 348.
34. Ibid., 347.
35. "Science and Religion," in *Adventurous Religion and Other Essays* (New York: Red Label Reprints, 1926), 92–93. Rather than assailing science for its metaphysical presumptions, Fosdick chastised those who used their own metaphysical assumptions to deny the validity of scientific discoveries: "There is no peace in sight between science and religion until religion recognizes that the sense of sanctity is too valuable an article to be misused in holding up scientific progress" (ibid., 103).
36. (New York: Macmillan, 1929), 123.
37. Ibid., 124–25.
38. "Science and Religion in the Nineteenth Century," in Needham, ed., *Science, Religion, and Reality*, 181.
39. Ibid., 182.
40. John Oman, "The Sphere of Religion," in Needham, ed., *Science, Religion, and Reality*, 261.
41. See Rudolf Otto, *The Idea of the Holy: An Inquiry into the Non-Rational Factor in the Idea of the Divine and Its Relation to the Rational*. 1923; trans. John W. Harvey (New York: Oxford University Press, 1958).
42. John Oman, "The Sphere of Religion," in Needham, ed., *Science, Religion, and Reality*, 283. For an analysis of the significance of Oman's theological work, see Stephen Bevans, *John Oman and His Doctrine of God* (Cambridge: Cambridge University Press, 1992).
43. Inge, "Conclusion, in Needham, ed.," *Science, Religion, and Reality*, 359–60.
44. Ibid., 360.
45. Ibid.
46. *Science and Human Life* (New York and London: Harper and Brothers, 1933), 133.
47. Ibid., 133–34.
48. *Science and the Modern World* (1925; rpt. New York: Macmillan, 1967), 189.
49. (New York: D. Appleton and Company, 1924). Matthews was dean of the Chicago Divinity School when he published this book.
50. Ibid., 351–422.
51. Ibid., 377.
52. For this discussion of personalism, I am indebted to Albert C. Knudson's

exhaustive account, *The Philosophy of Personalism: A Study in the Metaphysics of Religion* (New York: Abingdon, 1927). Knudson admits that the exact definition of "personality" had yet to be agreed upon.

53. Ibid., 237.
54. Ibid., 250.
55. "If both scientists and theologians had understood that science is by its very nature confined to the phenomenal realm and that religion is by its nature concerned simply with the ultimate power and purpose that lie back of phenomena, most of the conflicts between them in the past would have been avoided" (ibid., 253).
56. Eddington, *The Nature of the Physical World*, 275.
57. Ibid., 288.
58. Ibid., 333.
59. Eddington, *Science and the Unseen World*, 82.
60. "If it is difficult to separate out the subjective element in our knowledge of the external world, it must be much more difficult to distinguish it when we come to the problem of a self-knowing consciousness, where subject and object—that which knows and that which is known—are one and the same" (*Science and Religion: A Symposium*, 129).
61. J. S. Haldane, *Materialism* (London: Hodder and Stoughton, 1932), 131.
62. Ibid., 141.
63. Ibid., 218.
64. Ibid.
65. *Science and Religion: A Symposium*, 37.
66. Ibid., 43–44.
67. Ibid., 47.
68. Ibid.
69. Ibid.
70. Ibid., 53.
71. Julian Huxley, "Religion and Science," in *Essays of a Biologist* (1923; rpt. London: Chatto and Windus, 1928), 244–45.
72. Ibid., 284.
73. Ibid.
74. Ibid.
75. Julian Huxley's essay in *Science and Religion: A Symposium*, in Russell's view, contained "no support for even the most shadowy orthodoxy," while "it also contained little that liberal Churchmen would now find objectionable" (*R&S*, 173–74). Russell himself seemed to have found nothing worth further comment in Huxley's contribution, for there is not a single quotation from it in his notes on the book (Russell Archives, 210.0066y2-F5).
76. *Science and Religion: A Symposium*, 2–21. In his own essay in this anthology (83–92), H. R. L. Sheppard, as a theologian, disclaims scientific ability but asserts that the pursuit of truth is the primary goal of science. He disputes

Huxley's assertion that religion must assimilate the truths of science, by pointing out the wide disparity of scientific opinion on many subjects. He calls for coexistence, for more and better science and religion, and for less heat and more light in their controversies.

77. *Religion without Revelation* (London: Ernest Benn, 1927).
78. *Religion in an Age of Science* (New York: Frederick A. Stokes, 1929), 140–41.
79. *R&S*, 209.
80. Ibid., 7.
81. Ibid.
82. Ibid., 9.
83. Ibid., 36, 66, 105.
84. Ibid., 8.
85. He does note that the three elements of religion as a social phenomenon are not of equal weight: "A purely personal religion, so long as it is content to avoid assertions which science can disprove, may survive undisturbed in the most scientific age" (*R&S*, 9).
86. Ibid.
87. Ibid., 10.
88. Gifford Lectures, 1914 (London: Hodder and Stoughton, 1915).
89. *CP* 8, #10, 100. Russell also reviewed Balfour's second series from the newspaper reports (later published as *Theism and Thought* (1923)) and crisply commented, "It would seem that Lord Balfour, like many others, confuses science with the metaphysics of scientific men" (*CP* 9, #80, 441).
90. *CP* 8, #10, 101.
91. *CP* 8, 99, February 17, 1914.
92. *CP* 8, 99, February 24, 1914.
93. Russell Archives, 210.0066Y2-F5.
94. *R&S*, 174. Russell was asked to contribute to another BBC series. His lecture, "How Science has Changed Society," aired on January 6, 1932, and contained the same points of view reflected in *The Scientific Outlook* and *Religion and Science* (*CP* 10, #51, 395–402 and Appendix IV, 596–600). On the dilemma of the "old savage in the new civilization," he comments: "The next war, if it occurs, is therefore likely to prove far more destructive than what we still call the Great War. Scientific civilization cannot survive unless large-scale wars can be prevented" (398–99).
95. *Science and Religion: A Symposium* 24; *R&S*, 174–75.
96. *Science and Religion: A Symposium* 47; *R&S*, 175.
97. *Science and Religion: A Symposium* 76–77; *R&S*, 175. Russell was contemptuous of Malinowski's contribution, calling it a "pathetic avowal of a balked longing to belief in God" (*R&S*, 174).
98. Ibid., 176.
99. Ibid., 176–77.

100. Perhaps one reason for his rather genial discussion of this part of *Science and Religion: A Symposium* was due to his opinion of its main proponent; when Inge dismissed a religion based on the Victorian idea of universal progress through evolution, Russell concurred, saying, "on this matter . . . I find myself in agreement with the Dean, for whom, on many grounds, I have a very high respect" (ibid., 184).
101. *R&S,* 189.
102. Ibid.
103. Ibid., 17.
104. Ibid., 190.
105. Ibid. "I'm not quite sure what, but I believe it was scientific theologians and religiously-minded scientists."
106. Ibid., 191.
107. Ibid., 191–92. See, for example, C. Lloyd Morgan, *Emergent Evolution* (New York: Henry Holt, 1923) and Henri Bergson, *Creative Evolution,* trans. Arthur Mitchell (New York: Henry Holt and Company, 1911).
108. *R&S,* 194. "An omnipotent Being who created a world containing evil not due to sin must Himself be at least partially evil."
109. Ibid., 196–210.
110. Ibid., 213.
111. Ibid., 214.
112. Ibid., 216.
113. Ibid., 221–22.
114. Scholars have identified the negative effects of such an absolute dichotomy between fact and value. For example, "It is his fundamental pluralism—the splitting up of the world into a chain of unrelated facts—that we find a theory which offers little help in understanding either the world or society . . . What he wants with all his heart, the great humanitarian ideals he has worked for, are therefore completely out of reach" (John Lewis, *Bertrand Russell: Philosopher and Humanist* [New York: International Publishers, 1968], 84–85). Lewis presented an extreme picture, but there is some justice in the conclusion he reached: "All that Russell is so deeply concerned about socially and morally is left by him hanging in the air unrelated to actuality, lacking any principles of discrimination to discern truth from falsehood, right from wrong, unrelated to science, and not capable of being developed as a rational sociology. This is the inevitable consequence of the unrelieved dualism of fact and value, science and society, ethics and the natural order" (87–88).
115. *R&S* , 223.
116. Ibid.
117. Ibid., 230.
118. Ibid., 230–31.
119. Ibid., 232.

120. Ibid.
121. Ibid., 233–34.
122. Ibid., 236.
123. Ibid., 237.
124. Ibid., 242.
125. *AB*, 392–93. Benjamin R. Barber, in "Solipsistic Politics: Russell's Empiricist Liberalism," in George W. Roberts, ed., *Bertrand Russell Memorial Volume* (London: George Allen and Unwin, 1979), 455–78, writes about the effect of such solipsism later in Russell's life: "Russell . . . feels compelled by modern physics both to abandon all philosophical certitude about the world *and* to plunge nobly into the world to save it from the practical consequences of the self-same physics that has inspired his doubts" (465).
126. *ESO*, 145–46.
127. Ibid., 146.

CHAPTER 6

1. Marvin Kohl, "Bertrand Russell's Characterization of Benevolent Love," *Russell* n.s. 12, no. 2 (winter 1992–1993): 117–34.
2. *EGL*, 16.
3. Ibid., 26.
4. Ibid., 93.
5. Ibid., 122.
6. Ibid., 130.
7. Ibid., 135.
8. *M&M*, 134.
9. Ibid., 149.
10. Ibid., 241–42.
11. Ibid., 242.
12. Ibid., 244.
13. Ibid., 249.
14. Ibid., 233.
15. Ibid., 233–34.
16. *COH*, 21.
17. Ibid., 16.
18. Bol, "Mechanism and the Individual," 161–68.
19. *PIC*, 188. The three sources of power were military power, economic power, and power over opinion. The Russells believed that sociological development could be "controlled and completely changed by public opinion and the operation of human desires and beliefs" (ibid., 273).
20. *Icarus*, 57–61.
21. *Power*, 25.
22. Ibid., 9.

23. Ibid., 11.
24. "Power over men, not power over matter, is my theme in this book; but it is possible to establish a technological power over men which is based on power over matter" (ibid., 22).
25. Ibid., 23–24.
26. Ibid., 194.
27. See Freeden, *Liberalism Divided*, 319. He compares Russell to R. H. Tawney.
28. *Power*, 206.
29. Ibid., 81.
30. Ibid., 105.
31. Ibid., 107–08.
32. Ibid., 115–16.
33. Ibid., 172.
34. See Alan Ryan, *Russell: A Political Life* (1988; rpt. Harmondsworth, Middlesex: Penguin, 1990), 155; Bart Schultz, "Bertrand Russell in Ethics and Politics," *Ethics* 102 (April 1992), 594–634.
35. "The thesis of this book seems to me important, and I hoped it would attract more attention than it has done" (*AB*, 432).
36. "Bertrand Russell on Power," in *Ethics*, XLIX: 3 (April 1939): 253–85.
37. Ibid., 255.
38. Ibid., 256.
39. Ibid., 258, emphasis added.
40. Ibid., 264.
41. Ibid., 267 ff.
42. He continues: "No 'fact' of science is accepted as such unless competent observers agree about it. This means (a) that knowledge of agreement between minds, and hence validity of intercommunication of mental content, is prior to all knowledge of physical reality; and (b) that a judgment as to the competence of other knowers is also prior in validity to knowledge of concrete content" (ibid., 268–69).
43. Ibid., 156.
44. Ibid., 157.
45. Ibid., 172.
46. *Power*, 169.
47. *Outline*, 184.
48. Ibid., 187.
49. Ibid., 179.
50. Ibid., 185.
51. "Social cohesion demands a creed, or a code of behavior, or a prevailing sentiment, or, best, some combination of all three; without something of the kind, a community disintegrates" (*Power*, 104). Later in the book, he wrote: "To sum up: a creed or sentiment of some kind is essential to social cohe-

sion, but if it is to be a source of strength it must be genuinely and deeply felt by a considerable percentage of those upon whom technical efficiency depends" (105).
52. Ibid., 178.
53. "Controversies as to ends cannot be conducted, like scientific controversies, by appeals to facts; they must be conducted by an attempt to change men's feelings" (ibid., 170).
54. *FO*, 508.
55. Ibid.
56. *CP* 10 #34, 246–52.
57. Ibid., 247.
58. Ibid.
59. Ibid., 248.
60. While the problem of the existence of universals dogged Russell's footsteps throughout the interwar period, he did not reach a grudging acceptance of what this implied for his opposition to religion until the very last sentences of *An Inquiry into Meaning and Truth* (1940). Having spent the entire volume discussing the epistemological implications of language, only at the end does he admit the following: "We have arrived ... at a result which has been, in a sense, the goal of all our discussions. The result I have in mind is this: that complete metaphysical agnosticism is not compatible with the maintenance of linguistic propositions. Some modern philosophers hold that we know much about language, but nothing about anything else. This view forgets that language is an empirical phenomenon like another, and that a man who is metaphysically agnostic must deny that he knows when he uses a word. For my part, I believe that, partly by means of the study of syntax, we can arrive at considerable knowledge concerning the structure of the world" (*Inquiry*, 347).

Selected Bibliography

I. RUSSELL'S CONTEMPORARIES IN THE INTERWAR PERIOD

Ayres, C. E. "The Irony of Science: Can It Repair the Damage It Has Done?" *Forum* 86 (September 1931): 160–66.

Bakeless, John. *The Origin of the Next War: A Study in the Tensions of the Modern World.* New York: Viking, 1926.

Balfour, Arthur James. *Theism and Humanism.* Gifford Lectures, 1914. London: Hodder and Stoughton, 1915.

Barnes, Ernest William. *Scientific Theory and Religion.* New York: Macmillan, 1933.

Bergson, Henri. *Creative Evolution.* Translated by Arthur Mitchell. New York: Henry Holt and Company, 1911.

Briggs, Martin S. *Rusticus, or the Future of the Countryside.* London: K. Paul, Trench, Trubner, n.d.

Burtt, Edwin A. *Religion in an Age of Science.* New York: Frederick A. Stokes, 1929.

Chase, Stuart. *The Tyranny of Words.* New York: Harcourt Brace, 1938.

———. *Men and Machines.* 1929; rpt. New York: Macmillan, 1943.

Chisholm, Cecil. *Vulcan, or the Future of Labour.* London: K. Paul, Trench, Trubner, n.d.

Dearmer, Percy, ed. *Christianity and the Crisis.* London: Victor Gollancz, 1933.

Dewey, John. *The Quest for Certainty: A Study of the Relation between Knowledge and Action.* Gifford Lectures, 1929. London: George Allen and Unwin, 1930.

Durant, Will. *On the Meaning of Life.* New York: Ray Long and Richard Smith, 1932.

Eddington, Arthur. *Space, Time, and Gravitation: An Outline of the General Relativity Theory.* Cambridge: Cambridge University Press, 1921.

———. *Stars and Atoms*. New Haven, Conn.: Yale University Press, 1927.

———. *The Nature of the Physical World*. 1928; rpt. New York: Macmillan; Cambridge: Cambridge University Press, 1929.

———. *Science and the Unseen World*. Swarthmore Lecture, 1929. New York: Macmillan, 1929.

———. *The Expanding Universe*. Cambridge: Cambridge University Press, 1933.

———. *The Mathematical Theory of Relativity*. 1925; rpt. Cambridge: Cambridge University Press, 1937.

———. *New Pathways in Science*. 1934; rpt. Ann Arbor: University of Michigan Press, 1959.

———. *The Philosophy of the Physical Sciences*. 1938; rpt. Ann Arbor: University of Michigan Press, 1958.

Fosdick, Harry Emerson. *Adventurous Religion and Other Essays*. New York: Red Label Reprints, 1926.

———. *As I See Religion*. New York and London: Harper and Brothers, 1932.

Fosdick, Raymond. *The Old Savage in the New Civilization*. 1928; rpt. Garden City, N.Y.: Doubleday and Doran, 1929.

———. *The Story of the Rockefeller Foundation*. New York: Harper, 1952.

———. *Chronicle of a Generation: An Autobiography*. New York: Harper & Brothers, 1958.

Fournier D'Albe, E. E. *Quo Vadimus? Glimpses of the Future*. New York: Dutton, 1925.

———. *Hephaestus, the Soul of the Machine*. London: K. Paul, Trench, Trubner, n.d.

Fuller, J. F. C. *Pegasus, or Problems of Transport*. London: K. Paul, Trench, Trubner, 1926.

———. *Atlantis, or America and the Future*. London: K. Paul, Trench, Trubner, n.d.

Garrett, Garet. *Ouroboros, or the Mechanical Extension of Mankind*. London: Kegan Paul, Trench, Trubner, n.d.

Gloag, John. *Artifex, or the Future of Craftsmanship*. London: K. Paul, Trench, Trubner, n.d.

Haldane, J. B. S. *Daedalus, or Science and the Future*. 1924; rpt. New York: E. P. Dutton, 1925.

———. *Possible Worlds: A Scientist Looks at Science*. New York and London: Harper and Brothers, 1928.

———. *The Philosophical Basis of Biology*. Garden City, N.Y.: Doubleday and Doran, 1931.

———. *The Inequality of Man and Other Essays*. London: Chatto and Windus, 1932.

———. *Science and Human Life*. New York and London: Harper and Brothers, 1933.

———. *Science and Every Day Life*. N.p.: Lawrence and Wishart, 1939.

Haldane, J. S. *Materialism*. London: Hodder and Stoughton, 1932.

Hatfield, H. Stafford. *Automaton, or the Future of the Mechanical Man*. London: K. Paul, Trench, Trubner, 1925.

Huxley, Aldous. *Brave New World*. Chatto and Windus, 1932; rpt. London: Grafton Books, 1990.

Huxley, Julian. *Essays of a Biologist*. 1923; rpt. London: Chatto and Windus, 1928.

———. *Essays in Popular Science* 1926; rpt. London: Chatto and Windus, 1928.

———. *Religion without Revelation*. London: Ernest Benn, 1927.

———. *What Dare I Think? The Challenge of Modern Science to Human Action and Belief*. London: Chatto and Windus, 1931.

Inge, William Ralph. *Outspoken Essays*. First Series. 1919; rpt. London: Longmans, Green, 1923.

———. *The Idea of Progress*. Oxford: Clarendon, 1920.

———. *Outspoken Essays*. Second Series. London: Longmans, Green, 1922.

———. *Religion and Life: The Foundation of Personal Religion*. 1923; rpt. Freeport, N. J.: Books for Libraries Press, 1968.

———. *Lay Thoughts of a Dean*. Garden City, N.Y.: Garden City Publishing, 1926.

———. *The Church in the World: Collected Essays by William Ralph Inge*. London, New York: Longmans, Green & Co., 1928.

———. *The Social Teaching of the Church*. New York: Abingdon, 1930.

———. *Christian Ethics and Modern Problems*. New York and London: G. Putnam's Sons, 1930.

———. *More Lay Thoughts of a Dean*. New York: Putnam, 1931.

———. *Things New and Old*. Toronto: Longmans, Green & Co., 1933.

———. *God and the Astronomers*. London: Longmans, Green and Co., 1933.

———. *Vale*. London: Longmans, 1934.

Jacks, L. P. *Constructive Citizenship*. 1927; rpt. Garden City, N.Y.: Doubleday and Doran, 1928.

Jeans, James. *The Universe Around Us*. New York: Macmillan; Cambridge: Cambridge University Press, 1929.

———. *The Mysterious Universe*. New York: Macmillan; Cambridge: Cambridge University Press, 1930.

———. *The Stars in Their Courses*. New York: Macmillan; Cambridge: Cambridge University Press, 1931.

———. *The New Background of Science*. Cambridge: Cambridge University Press, 1933.

———. *Through Space and Time*. New York: Macmillan; Cambridge: Cambridge University Press, 1934.

Joad, C. E. M. *Essays in Common Sense Philosophy*. London: Headley Brothers, 1919.

———. *Thrasymachus, or the Future of Morals*. New York: Dutton, 1926.

———. *Matter, Life, and Value*. Oxford: Oxford University Press, 1929.

———. *Philosophical Aspects of Modern Science*. London: George Allen and Unwin, 1932.

———. *The Future of Life: A Theory of Vitalism*. London and New York: G. P. Putnam's Sons, 1928.

———. *The Book of Joad: A Belligerent Autobiography*. 1932; rpt. London: Faber and Faber, 1943.

Kallet, Arthur, and F. J. Schlink. *100,000,000 Guinea Pigs: Dangers in Everyday Drugs and Cosmetics*. New York: Vanguard Press, 1933.

Kenworthy, J. M. *New Wars, New Weapons. A Library of New Ideas*, No. 1. London: Elkin Matthews and Marrot, 1930.

Knudson, Albert C. *The Philosophy of Personalism: A Study in the Metaphysics of Religion*. New York: Abingdon, 1927.

Lardner, John, and Thomas Sugrue. *The Crowning of Technocracy*. New York: Robert M. McBride, 1933.

Liddell Hart, B. H. *Paris, or the Future of War*. New York: Dutton, 1925.

Lippmann, Walter. *A Preface to Morals*. New York: Macmillan, 1929.

Lynd, Robert S., and Helen Merrell Lynd. *Middletown: A Study in American Culture*. New York: Harcourt, Brace, 1929.

Matthews, Shailer. *Contributions of Science to Religion*. New York: D. Appleton and Company, 1924.

McDougall, William. *World Chaos: The Responsibility of Science*. London: Kegan Paul, Trench, Trubner and Co., 1931.

———. *Janus: The Conquest of War: A Psychological Inquiry*. 1927; rpt. New York and London: Garland Publishing, 1972.

Millikan, Robert A. *Evolution in Science and Religion*. New Haven, Conn.: Yale University Press, 1927; rpt. "with slight additions," 1935.

———. *Science and the New Civilization*. New York: Charles Scribner's Sons, 1930.

Morgan, C. Lloyd. *Emergent Evolution*. New York: Henry Holt, 1923.

Mumford, Lewis. *Technics and Civilization*. New York: Harcourt, Brace and Company, 1934.

Needham, Joseph, ed. *Science, Religion, and Reality*. New York: Macmillan, 1925; rpt. New York: George Braziller, 1955.

———. *The Skeptical Biologist*. London: Chatto and Windus, 1929.

Ogden, C. K., and I. A. Richards. *The Meaning of Meaning: A Study of the Influence of Language upon Thought and of the Science of Symbolism*. 1923; rpt. San Diego: Harcourt, Brace Jovanovich, 1989.

Phillips, Thomas R., "Debunking Mars' Newest Toy." *Saturday Evening Post*, March 4, 1933.

Pupin, Michael. *From Immigrant to Inventor*. New York: Charles Scribner's Sons, 1927.

———. *The New Reformation: From Physical to Spiritual Realities*. New York and London: Charles Scribner's Sons, 1929.

———. *Romance of the Machine*. New York: Charles Scribner's Sons, 1930.

Russell, Dora. *The Right to be Happy*. New York and London: Harper & Brothers, 1927.

Santayana, George. *Scepticism and Animal Faith: Introduction to a System of Philosophy*. 1923; rpt. N.p.: Dover, 1955.

Schiller, F. C. S. *Tantalus, or the Future of Man*. 1924; rpt. New York: E. P. Dutton, 1925.

Schweitzer, Albert. *Out of My Life and Thought: An Autobiography*. Translated by C. T. Campion. New York: Henry Holt, 1933.

———. *The Philosophy of Civilization*. Translated by C. T. Campion. New York: Macmillan, 1949.

Science and Religion: A Symposium. London: Gerald Howe and New York: Charles Scribner's Sons, 1931.

Slosson, E. E. *Keeping Up with Science*. New York: Harcourt Brace, 1924.

———, and Otis Caldwell. *Science Remaking the World*. Garden City, N.Y.: Garden City Publishing, 1925.

Smuts, J. C. *Holism and Evolution*. New York: Macmillan, 1926.

Streeter, Burnett Hillman. *Reality: A New Correlation of Science and Religion*. London: Macmillan, 1927.

Sullivan, J. W. N. *Gallio or the Tyranny of Science*. London: K. Paul, Trench, Trubner, n.d.

———. *Limitations of Science.*1933; rpt. Harmondsworth, Middlesex: Penguin, 1938.

Thomson, J. Arthur. *The Control of Life.* New York: Henry Holt, 1921.

———. *What is Man?* London: Methuen, 1923.

———. *Science and Religion.* New York: Charles Scribner's Sons, 1925.

Veblen, Thorstein. *The Place of Science in Modern Civilization and Other Essays.* New York: B. W. Huebsch, 1919.

Whitehead, Alfred North. *Science and the Modern World.* 1925; rpt. New York: Macmillan, 1967.

Wiggam, Albert Edward. *The New Decalogue of Science.* Garden City, N.Y.: Garden City Publishing, n.d.

II. SELECTED SECONDARY AND OTHER SOURCES

Aiken, Lillian. *Bertrand Russell's Philosophy of Morals.* New York: Humanities Press, 1963.

Andersson, Stefan. *In Quest of Certainty: Bertrand Russell's search for certainty in religion and mathematics up to* The Principles of Mathematics *(1903).* Ph.D. dissertation. Stockholm: Almqvist & Wiksell, 1994.

———. "Religion in the Russell Family." *Russell* n.s., 13, no. 2 (winter 1993–1994): 117–49.

Ayer, A. J. *Bertrand Russell.* New York: Viking, 1972.

Bevans, Stephen. *John Oman and His Doctrine of God.* Cambridge: Cambridge University Press, 1992.

Blackwell, Kenneth. *The Spinozistic Ethics of Bertrand Russell.* London: Allen and Unwin, 1985.

———, and Harry Ruja. *A Bibliography of Bertrand Russell, Volume 2, Serial Publications 1890–1990.* New York and London: Routledge, 1995.

Brink, Andrew. "Bertrand Russell and the Decline of Mysticism." *Russell* n.s., 7, no. 1 (summer 1987): 42–52.

———. *Bertrand Russell: The Psychobiography of a Moralist.* Atlantic Highlands, N.J.: Humanities Press, 1989.

Brooke, John Hedley. *Science and Religion: Some Historical Perspectives.* Cambridge: Cambridge University Press, 1991.

———, and Geoffrey Cantor. *Reconstructing Nature: The Engagement of Science and Religion.* Gifford Lectures 1995–1996. Edinburgh: T & T Clark, 1998.

Clark, Ronald. *The Life of Bertrand Russell.* London: J. Cape, 1975.

Douglas, A. Vibert. *The Life of Arthur Stanley Eddington.* London: Thomas Nelson and Sons, 1957.

Draper, J. W. *History of the Conflict between Religion and Science.* 1874; rpt. New York: D. Appleton and Company, 1897.

Duran, Jane. "Russell on Pragmatism." *Russell* n.s., 14, no. 1 (summer 1994): 31–38.

Eames, Elizabeth Ramsden. *Bertrand Russell's Theory of Knowledge.* London: George Allen and Unwin, 1969.

———. *Bertrand Russell's Dialogue with his Contemporaries.* Carbondale and Edwardsville: Southern Illinois University Press, 1989.

Freeden, Michael. *Liberalism Divided: A Study in British Political Thought.* Oxford: Clarendon, 1986.

Greenspan, Louis. *The Incompatible Prophesies: Bertrand Russell on Science and Liberty.* Oakville: Mosaic Press, 1978.

———, and Stefan Andersson, eds. *Russell on Religion.* London and New York: Routledge, 1999.

Griffin, Nicholas. *Russell's Idealist Apprenticeship.* Oxford: Clarendon, 1991.

———. "The Legacy of Russell's Idealism for His Later Philosophy." *Russell* n.s., 12, no. 2 (winter 1992–1993): 117–34.

———. "Russell as a Critic of Religion." *Studies in Religion* 24:1 (1995): 47–58.

Hagen, Paul. "Why Russell Didn't Think He Was a Philosopher of Education." *Russell* n.s., 13, no. 2 (winter 1993–1994): 150–67.

Harrison, Royden. "Bertrand Russell: From liberalism to socialism?" *Russell* n.s., 6, no. 1 (summer 1986): 5–38.

Hylton, Peter. *Russell, Idealism, and the Emergence of Analytic Philosophy.* Oxford: Clarendon, 1990.

Jager, Ronald. *The Development of Bertrand Russell's Philosophy.* London: George Allen and Unwin, 1972.

Kohl, Marvin. "Bertrand Russell's Characterization of Benevolent Love," *Russell* n.s., 12, no. 2 (winter 1992–1993): 117–34.

Lewis, John. *Bertrand Russell: Philosopher and Humanist.* New York: International Publishers, 1968.

Lindberg, David C., and Ronald L. Numbers. "Beyond War and Peace: A Reappraisal of the Encounter between Christianity and Science." *Church History* 55: 3 (September 1986): 338–54.

Lippincott, Mark S. "Russell's leviathan." *Russell* n.s., 10, no. 1 (summer 1990): 6–29.

Monk, Ray. *Bertrand Russell: The Spirit of Solitude.* London: Jonathan Cape, 1996.

Moore, James. *The Post–Darwinian Controversies: A Study of the Protestant*

Struggle to Come to Terms with Darwin in Great Britain and America, 1870–1900. Cambridge: Cambridge University Press, 1979.

———. "Speaking of 'Science and Religion'—Then and Now." *History of Science* XXX (1992): 311–23.

Moorehead, Caroline. *Bertrand Russell.* London: Sinclair-Stevenson, 1992.

Moran, Margaret, and Carl Spadoni, eds. *Intellect and Social Conscience: Essays on Bertrand Russell's Early Work.* Hamilton: McMaster University Library Press, 1984.

Morrell, Jack, and Arnold Thackray. *Gentlemen of Science.* Oxford: Clarendon, 1981.

Nathanson, Stephen. "Russell's scientific mysticism." *Russell* n.s., 5, no. 1 (summer 1985): 14–25.

Pears, D. F. *Bertrand Russell and the British Tradition in Philosophy.* New York: Random House, 1967.

Revoldt, Daryl L. *Raymond B. Fosdick: Reform, Internationalism and the Rockefeller Foundation.* Ph.D. dissertation. University of Akron, 1982.

Roberts, George W., ed. *Bertrand Russell Memorial Volume.* London: George Allen and Unwin, 1979.

Rockler, Michael J. "The Curricular Role of Russell's Scepticism." *Russell* n.s., 12, no. 1 (summer 1992): 50–60.

Ryan, Alan. *Russell: A Political Life.* 1988; rpt. Harmondsworth, Middlesex: Penguin, 1990.

Schilpp, Paul, ed. *The Philosophy of Bertrand Russell.* New York: Tudor, 1951.

Schultz, Bart. "Bertrand Russell in Ethics and Politics." *Ethics* 102 (April 1992): 594–634.

Slater, John G. *Bertrand Russell.* Bristol: Thoemmes Press, 1994.

Thomas, J. E., and Ken Blackwell. *Russell in Review: The Bertrand Russell Centenary Celebrations.* Toronto: Samuel Stevens, Hakkert, 1976.

Tully, R. E. "Three Studies of Russell's Neutral Monism." *Russell* n.s., 13, no. 1 (summer 1993): 5–35; n.s., 13, no. 2 (winter 1993–1994): 185–202.

Vellacott, Jo. *Bertrand Russell and the Pacifists in the First World War.* Brighton, Sussex: Harvester Press, 1980.

White, Andrew Dickson. *A History of the Warfare of Science with Theology in Christendom.* 2 vols. 1896; rpt. New York and London: D. Appleton, 1926.

Wilson, David B. "On the Importance of Eliminating Science and Religion from the History of Science and Religion: The Cases of Oliver Lodge, J. H. Jeans and A. S. Eddington," in *Facets of Faith and Science,* edited by Jitse van der Meer, vol. 2. Lanham, Md.: University Press of America, 1996.

Winchester, Ian, and Ken Blackwell, eds. *Antimonies and Paradoxes: Studies in Russell's Early Philosophy*. Hamilton: McMaster University Library Press, 1989.

Wood, Alan. *Bertrand Russell: The Passionate Sceptic*. London: George Allen and Unwin, 1957.

Woodhouse, Howard. "Science as method: the conceptual link between Russell's philosophy and his educational thought." *Russell* n.s., 5, no. 2 (winter 1985–1986): 150–61.

Index

Alexander, Samuel, 106
Aliotta, Antonio, 94, 152n20
Andersson, Stefan, 141n13
Archimedes, 149n108
Ayer, A. J., 142n45
Ayres, C. E., 3–4

Bakeless, John, 129n1
Balfour, A. J., 91, 92–93, 102, 151n14, 152n18
Barber, Benjamin R., 157n125
Barnes, E. W., Bishop of Birmingham, 92, 106
Bedford, A. C., 137n34
Bentham, Jeremy, 54
Bergson, Henri, 56, 106, 156n107
Bevans, Stephen, 153n42
Bohr, Nils, 144n3
Booth, Wayne C., 130n17
Briggs, Martin S., 133n25
Brightman, Edgar, 45, 48
British Association for the Advancement of Science, 1, 85
British Broadcasting Corporation (BBC), xxv, 91, 103, 130n20, 155n94
Broad, C. H., 144n1
Brooke, John Hedley, 150n1
Brown, William, 152n20
Bruno, Giordano, 88
Buddha, 43
Burroughs, E. A., Bishop of Ripon, 1, 11–13, 53, 142n53
Burtt, Edwin A., 99

Cantor, Geoffrey, 150n1
Capitalism, 24

Chase, Stuart, 4
Churchill, Winston, 140n8
Coleridge, Samuel Taylor, 85
Cooper Union Hall, 88
Copernicus, Nicholas, 88
Cornell University, 88

Darwin, Charles, 89, 149n108
Dirac, Paul, 144n3
Draper, J. W., xxvi, 87–88, 90, 92
Dreiser, Theodore, 42
Durant, Will, 41–42

Eddington, Arthur, xxv, 39, 60–62, 68, 72–82, 91, 97, 104, 136n78, 144n1, 144n2, 146n62, 147n63, 148n85, 150n112, 152n21
Edison, Thomas, 140n8
Einstein, Albert, xxi, 60, 65, 132n7, 144n3
Ellis, Havelock, 42
Ethics, social, xxiii; and religion, xxiv

Ford, Henry, xiv, 5, 6, 133n19
Fordism, 5–6
Fosdick, Harry Emerson, 8, 93, 134n37
Fosdick, Raymond Blaine, 2, 7–13
Fournier D'Albe, E. E., 133n30
Freeden, Michael, 138n43
Freud, Sigmund, 69
Fuller, J. F. C., 5–6, 129n1, 133n20

Galileo, 88–89, 103, 149n108, 151n15
Garrett, Garet, 133n25
Gloag, John, 133n25
Great War, The (1914–1918), effects of, xi–xii, xv, xxii, 1, 18, 21

Greenspan, Louis, xviii, 141n13
Griffin, Nicholas, xix, 141n13

Haldane, J. B. S., 3, 61–62, 95, 131n4, 132n7
Haldane, J. S., 97–98, 100, 103–4, 106
Haldemann-Julius, E., xvii, 50
Hatfield, H. Stafford, 134n30
Heisenberg, Werner, 144n4
Hibbert Journal, 55
Hume, David, 70
Huxley, Aldous, xxiii, 5, 31, 122
Huxley, Julian, 98–99, 103, 154nn75–76

Industrial civilization, nature of, xiv, xxiii, 10, 18–22, 24, 111, and morality, 113–14; power in, 26; prospects of, 2, 5, 8, 13, 26–27
Inge, William Ralph, Dean of St. Paul's, 91, 93, 95, 104, 140n8, 151n14, 152n19
Instrumentalism, xiv, 111
Internationalism, 23,
Interwar period, nature of, xii

James, William, 139n82, 141n12
Jeans, James, xvii, xxv, 39, 61–62, 76–79, 81–82, 106, 144n2
Joad, C. E. M., 52–53, 129n1, 150n112

Kepler, Johannes, 88, 149n108
Kevles, Daniel, 131n1, 152n16
Knight, Frank H., 117–18
Knudson, Albert C., 153n52
Kohl, Marvin, 157n1

Lao Tse, 43
Lewis, John, 156n114
Liddell Hart, B. H., 129n1
Lindberg, David C., 151n4
Lindemann, Eduard, xviii
Lippmann, Walter, 93–94
Locke, John, 116
Lucretius, 45

Machine Age, 4–5, 7, 12
Machine civilization, 6, 9
Malinowski, Bronislaw, 91, 104
Malthus, 12

Marconi, Guglielmo, 140n8
Marx, Karl, 115
Matthews, Shailer, 96
McDougall, William, 7, 12–14
Metaphysics, definition of, xx
Meyer, Ben, 130n17
Millikan, Robert A., 3, 7, 11–13, 132n7
Modern, definition of, xix
Montesquieu, 116
Moore, James, 150nn1, 3, 151nn13, 15
Moral development, need for, 2
Morgan, C. Lloyd, 106, 156n107
Morrell, Ottoline, 43, 47, 102
Muggeridge, Malcolm, 42
Mumford, Lewis, xix, 4
Mussolini, Benito, 140n8
Mysticism, see Russell, Bertrand, on religious experience

Nationalism, 22
Needham, Joseph, xxv, 152nn17, 20
Nehru, Jawaharlal, 42
Newton, Isaac, 86, 149n108
Next War, The, fear of, xi–xii, 1–2, 16, 19; preventing, xiii
Niebuhr, Reinhold, 53
Numbers, Ronald, 151n4

Oman, John, 94, 152n20
Operational metaphysic, and Bertrand Russell, 120–23
Otto, Rudolph, 153n41
Outlook, mechanistic, xiv, xxii, 20
Outlook, scientific, xxi, xxiii, 27–28, 36–40, 42, 51, 58, 108, 111
Outlook, utilitarian, 32–33

Pavlov, 70, 143n68, 149n108
Pearsall, Alys, 141n11
Performance, rhetorical, xv–xvii
Phillips, Thomas R., 132n8
Physics, philosophical implications of, xxi, xxiv, 14–16, 60–63, 72, 78–81; and matter, 63–65; and causality, 65–66; and mind/matter duality, 67–68; according to Eddington, 72–76; according to Jeans, 76–78
Plato, 53

Popular literature, definition of, xix
Pupin, Michael, 6
Pythagoras, retreat from, Russell's, xiii, xv, 43, 63, 108, 124, 129n9, 141n13

Religion and Science, see Science and Religion
Revoldt, Daryl, 134n35
Richards, I. A., 135n72
Rockefeller Foundation, and Fosdick, Harry Emerson, 8; and Fosdick, Raymond Blaine, 8
Russell, Bertrand, on cosmic purpose, 105–6; on Eddington and Jeans, 60–62, 72–79; on education, 113; on the good life, 112; on knowledge, xviii; on liberal ideals, 112–13; on religion, 42–46; on religion and ethics, 52–54; on religion, as social institution, 49–51, 59; on religious experience, 46–48, 55–58, 103–5; on science, conception of, xvii–xviii, xx, 81–82; on science and ethics, 107–8, 118–19; on science and religion, 100–105, 121;
  AB, xv, 129n7, 136n13, 140n8, 141n10, 157n125, 158n35
  ABCA, xxiv, 65
  ABCRel, xxiv, 64–65
  AMatter, xxiv
  AMind, xxiv, 67, 146n34
  BOL, 26, 50, 115
  COH, 114–15
  CP9, 140nn98, 99
  CP10, xvii, 77, 94, 121, 129n12, 142nn36, 39, 18nn93, 94, 95, 155nn89–92, 158 nn56–59
  CP12, 47–48
  EGL, 113
  ESO, 157nn126, 127
  FO, 18, 121
  HWP, 129n9
  Icarus, 27, 115, 132n4
  Inquiry, 122–23, 159n60
  M&M, 113–14
  ML, 46–47, 55–57
  MPD, xii–xiii
  OKEW 60, 144n1
  Outline, 60, 62–68, 144n9, 145nn14, 22, 29, 146n30
  PFM, xii–xiii
  PI, 25, 32
  PIC, xiii, 19–20, 21–24, 26, 32–33, 115, 120, 157n19
  Power, xxvi, 115–20, 123, 158n51
  PPhil, xv, 61
  PRF, 25
  PSR, 20–21, 50–51, 54
  R&S, xv–xvi, xxv–xxvi, 44, 83–84, 100–109, 120
  SceptEss, 144nn6, 7
  SO, xxiii, 27–40, 61–62, 68–72, 74, 79–81, 120, 146n46, 148n84, 149nn98, 101, 103, 108
  WIB, 53–54, 132n4
  WNC, 45–46, 49, 52
Russell, Colin, 150n1
Russell, Dora, 136n5
Ryan, Alan, 158n34

Santayana, George, 71, 146n54
Sarton, George, 151n14
Schiller, F. C. S., 2, 131n3, 132n4, 139n82
Schilpp, Paul, xvi
Schultz, Bart, 158n34
Schweitzer, Albert, xxiii, 36–38, 53, 75, 122–24
Science and Religion, conflict scenario, xxv–xxvi, 87–89; in interwar period, 89–96; and personalism, 96–99; usage, xix–xx, 83–87
Science and Religion: A Symposium, xxv, 91–92, 96–99, 103, 105
Science, and industrialism, 3, 26; and knowledge, 68–71; and power, 10, 110, 113, 115–17; and society, 3, 11, 19, 29–31, 109, 114; and technology, 4, 7, 13, 14, 26; and values, 34–35
Science, Religion, and Reality, xxv, 91–95
Scott, Howard, 132n18
Sheppard, H. R. L., 154n76
Singer, Charles, 152n20
Slater, John, 142n35
Slosson, E. E., 89
Socialism, 25

Stalin, Josef, 140n8
Stravinsky, Igor, 140n8
Streeter, B. H., 103
Sullivan, J. W. N., 7, 14–17

Tawney, R. H., 158n27
Technique, definition of, xix
Technocracy, 5
Thomson, J. Arthur, 103
Thrasymachus, 53
Tiergarten programme, xix
Today and Tomorrow series, 2

Veblen, Thorstein, 136n12
Vesalius, Andreas, 88

Watson, James, 143n68
Webb, Clement C., 152n20
Whewell, William, 85
White, Andrew Dickson, 83, 88–90, 101, 104
Whitehead, Alfred North, 95, 144n1
Wilson, David B., 150nn1, 3

Zalowitz, Nathaniel, xvii